Joseph Henry Wythe

The Microscopist

A Manual of Microscopy and Compendium of the Microscopic Science

Joseph Henry Wythe

The Microscopist
A Manual of Microscopy and Compendium of the Microscopic Science

ISBN/EAN: 9783337404130

Printed in Europe, USA, Canada, Australia, Japan

Cover: Foto ©berggeist007 / pixelio.de

More available books at **www.hansebooks.com**

THE MICROSCOPIST.

ZENTMAYER'S LARGEST MICROSCOPE.
ONE-THIRD ACTUAL SIZE.

THE MICROSCOPIST:

A MANUAL OF MICROSCOPY

AND

COMPENDIUM OF THE MICROSCOPIC SCIENCES,

MICRO-MINERALOGY, MICRO-CHEMISTRY, BIOLOGY, HISTOLOGY,
AND PATHOLOGICAL HISTOLOGY.

THIRD EDITION.

REWRITTEN AND GREATLY ENLARGED.

WITH

TWO HUNDRED AND FIVE ILLUSTRATIONS.

BY

J. H. WYTHE, A.M., M.D.,

PROFESSOR OF MICROSCOPY AND BIOLOGY IN THE MEDICAL COLLEGE OF THE PACIFIC,
SAN FRANCISCO.

RESPECTFULLY DEDICATED

TO THE

SAN FRANCISCO MICROSCOPICAL SOCIETY,

AS A TESTIMONY

TO THE

ZEAL AND INDUSTRY OF ITS MEMBERS

IN THE PROSECUTION

OF

MICROSCOPIC SCIENCE.

PREFACE.

The progress of microscopic science may be well illustrated by a comparison between the present and former editions of this book. The author's intention was to place within the reach of the student of nature a compendium of microscopy, free from unnecessary verbiage, which should aid in every department of natural science. It is no small compliment to such a work that for a quarter of a century it should hold a place among works of reference, although surrounded by larger and more pretentious volumes. In order to meet the request of the publishers for another edition, it has been found necessary to rewrite the entire book, and although the original design has been kept in view, the numerous additions to our science render considerable enlargement needful, notwithstanding the effort made to concentrate the material into the smallest compass consistent with perspicuity.

The vision of microscopy sweeps over all the world, and embraces all forms of organic and inorganic existence. To give directions respecting most approved methods, and to classify the most important facts, has required labor, which it is hoped will result in rendering the work a necessary companion to the student, and an aid to the progress of real science.

Many of the figures illustrating the lower forms of life, and normal and pathological histology, have been drawn from the works of Carpenter, Beale, Frey, Stricker, Billroth, and Rindfleisch, to which the more advanced student is referred for further details.

January, 1877.

CONTENTS.

CHAPTER I.

HISTORY AND IMPORTANCE OF MICROSCOPY.

Application of the Microscope to Science and Art—Progress of Microscopy, ... 17–21

CHAPTER II.

THE MICROSCOPE.

The Simple Microscope—Chromatic and Spherical Aberration—Compound Microscope—Achromatic Object-glasses—Eye-pieces—Mechanical Arrangements—Binocular Microscope, . . . 21–32

CHAPTER III.

MICROSCOPIC ACCESSORIES.

Diaphragms—Condensers—Oblique Illuminators—Dark-ground ditto—Illumination of Opaque Objects—Measuring and Drawing Objects—Standards of Measurement—Moist Chamber—Gas Chamber—Warm Stage—Polariscope—Microspectroscope—Nose-piece—Object-finders—Micro-photography, ... 32–48

CHAPTER IV.

USE OF THE MICROSCOPE.

Care of the Instrument—Care of the Eyes—Table—Light—Adjustments—Errors of Interpretation—Testing the Microscope, . . 48–58

CHAPTER V.

Modern Methods of Examination.

Preliminary Preparation of Objects—Minute Dissection—Preparation of Loose Textures—Preparation by Teasing—Preparation by Section—Staining Tissues—Injecting Tissues—Preparation in Viscid Media—Fluid Media—Indifferent Fluids—Chemical Reagents—Staining Fluids—Injecting Fluids—Preservative Fluids—Cements, . 58–76

CHAPTER VI.

Mounting and Preserving Microscopic Objects.

Opaque Objects—Cells—Dry Objects—Mounting in Balsam or Dammar—Mounting in Fluid—Cabinets—Collecting Objects—Aquaria, 76–83

CHAPTER VII.

The Microscope in Mineralogy and Geology.

Preparation of Specimens—Examination of Specimens—Crystalline Forms—Crystals within Crystals—Cavities in Crystals—Use of Polarized Light—Origin of Rock Specimens—Materials of Organic Origin—Microscopic Palæontology, ... 84–98

CHAPTER VIII.

The Microscope in Chemistry.

Apparatus and Modes of Investigation—Preparation of Crystals for the Polariscope—Use of the Microspectroscope—Inverted Microscope—General Micro-chemical Tests—Determination of Substances—Alkalies—Acids—Metallic Oxides—Alkaloids—Crystalline Forms of Salts, ... 98–115

CHAPTER IX.

The Microscope in Biology.

Theories of Life—Elementary Unit or Cell—Cell-structure and Formation—Phenomena of Bioplasm—Movements of Cells—Microscopic Demonstration of Bioplasm—Chemistry of Cells and their Products—Varieties of Bioplasm—Cell-genesis—Reproduction in Higher Organisms—Alternation of Generations—Parthenogenesis—Transformation and Metamorphosis—Discrimination of Living Forms, 116–127

CHAPTER X.

THE MICROSCOPE IN VEGETABLE HISTOLOGY AND BOTANY.

Molecular Coalescence—Cell-substance in Vegetables—Cell-wall or Membrane—Ligneous Tissue—Spiral Vessels—Laticiferous Vessels—Siliceous Structures—Formed Material in Cells—Forms of Vegetable Cells—Botanical Arrangement of Plants—Fungi—Protophytes—Desmids—Diatoms—Nostoc—Oscillatoria—Examination of the Higher Cryptogamia—Examination of Higher Plants, . 128–157

CHAPTER XI.

THE MICROSCOPE IN ZOOLOGY.

Monera—Rhizopods—Infusoria—Rotatoria—Polyps—Hydroids—Acalephs—Echinoderms—Bryozoa—Tunicata—Conchifera—Gasteropoda—Cephalopoda—Entozoa—Annulata—Crustacea—Insects—Arachnida—Classification of the Invertebrata, . . . 158–182

CHAPTER XII.

THE MICROSCOPE IN ANIMAL HISTOLOGY.

Histo-chemistry—Histological Structure—Simple Tissues—Blood—Lymph and Chyle—Mucus—Epithelium—Hair and Nails—Enamel—Connective Tissues—Compound Tissues—Muscle—Nerve—Glandular and Vascular Tissue—Development of the Tissues—Digestive and Circulatory Organs—Secretive Organs—Respiratory Organs—Generative Organs—Locomotive Organs—Sensory Organs—Organs of Special Sense—Suggestions for Practice, . . . 182–226

CHAPTER XIII.

THE MICROSCOPE IN PRACTICAL MEDICINE AND PATHOLOGY.

Microscopic Appearances after Death of the Tissues—Morbid Action in Tissues—New Formations—Examination of Urinary Deposits—Human Parasites—Examination of Sputa—Microscopic Hints in Materia Medica and Pharmacy, ... 226–245

THE MICROSCOPIST.

CHAPTER I.

HISTORY AND IMPORTANCE OF MICROSCOPY.

The term microscopy, meaning the use of the microscope, is also applied to the knowledge obtained by this instrument, and in this sense is commensurate with a knowledge of the minute structure of the universe, so far as it may come under human observation. Physics and astronomy treat of the general arrangement and motions of masses of matter, chemistry investigates their constitution, and microscopy determines their minute structure. The science of histology, so important to anatomy and physiology, is wholly the product of microscopy, while this latter subject lends its aid to almost every other branch of natural science.

To the student of physical phenomena this subject unfolds an amazing variety developed from most simple beginnings, while to the Christian philosopher it gives the clearest evidence of that Creative Power and Wisdom before whom great and small are terms without meaning.

In the arts, as well as in scientific investigations, the microscope is used for the examination and preparation of delicate work. The jeweller, the engraver, and the miner find a simple microscope almost essential to their employments. This application of the magnifying power of lenses was known to the ancients, as is shown by the glass lens found at Nineveh, and by the numerous gems and tablets so finely engraved as to need a magnifying glass to detect their details.

In commerce, the microscope has been used to detect adulterations in articles of food, drugs, and manufactures. In a single year $60,000 worth of adulterated drugs was condemned by the New York inspector, and, so long as selfishness is an attribute of degraded humanity, so long will the microscope be needed in this department.

In agriculture and horticulture microscopy affords valuable assistance. It has shown us that mildew and rust in wheat and other food-grains, the "potato disease," and the "vine disease," are dependent on the growth of minute parasitic fungi. It has also revealed many of the minute insects which prey upon our grain-bearing plants and fruit trees. The damage wrought by these insects in the United States alone has been estimated by competent observers as not less than three hundred millions of dollars in each year. The muscardine, which destroys such large numbers of silk-worms in France and other places, is caused by a microscopic fungus, the *Botrytis bassiana*.

The mineralogist determines the character of minute specimens or of thin sections of rock, and the geologist finds the nature of many fossil remains by their magnified image in the microscope.

The chemist recognizes with this instrument excessively minute quantities and reactions which would otherwise escape observation. Dr. Wormley shows that micro-chemical analysis detects the reaction of the 10,000th to the 100,000th part of a grain of hydrocyanic acid, mercury, or arsenic, and very minute quantities of the vegetable alkaloids may be known by a magnified view of their sublimates. The micro-spectroscope promises still more wonderful powers of analysis by the investigation of the absorption bands in the spectra of different substances.

In biology the wonderful powers of the microscope find

their widest range. If we see not life itself, we see its first beginnings, and the process of its development or manifestation. If we see not Nature in her undress, we trace the elementary warp and woof of her mystic drapery.

In vegetable and animal physiology we see, by its means, not only the elementary unit—the foundation-stone of the building—but also chambers and laboratories in the animated temple, which we should never have suspected—tissues and structures not otherwise discoverable —not to speak of species innumerable which are invisible to the naked eye.

In medical science and jurisprudence the contributions of microscopy have been so numerous that constant study in this department is needed by the physician who would excel or even keep pace with the progress of his profession. Microscopy may be truly called the guiding genius of medical science.

Even theology has its contribution from microscopy. The teleological view of nature, which traces design, receives from it a multitude of illustrations. In this department the war between skeptical philosophy and theology has waged most fiercely; and if the difference between living and non-living matter may be demonstrated by the microscope, as argued by Dr. Beale and others, theology sends forth a pæan of victory from the battlements of this science.

The attempts made by early microscopists to determine ultimate structure were of but little value from the imperfections of the instruments employed, the natural mistakes made in judging the novel appearances presented, and the treatment to which preparations were subjected. In late years the optical and mechanical improvements in microscopes have removed one source of error, but other sources still remain, rendering careful attention to details and accurate judgment of phenomena quite essential. Careful manipulation and minute dissection require a knowledge of the effects of various physical and chemical agencies, a steady hand, and a quick-discerning eye. Above all, microscopy requires a cultured mind, capable of readily detecting sources of fallacy, and such a love of truth as enables a man to free himself from all preconceived notions of structure and from all bias in favor of particular theories and analogies. What result is it possible to draw from the observations of those who boil, roast, macerate, putrefy, triturate, and otherwise injure delicate tissues, except for the purpose of isolating special structures or learning the effects of such agencies? Yet many of the phenomena resulting from such measures have been described as primary, and theories of development have been proposed on the basis of such imperfect knowledge.

Borelli (1608–1656) is considered to be the first who applied the microscope to the examination of animal structure. Malpighi (1661) first witnessed the actual circulation of the blood, which demonstrated the truth of Harvey's reasoning. He also made many accurate observations in minute anatomy. Lewenhoeck, Swammerdam, Lyonet, Lieberkuhn, Hewson, and others, labored also in this department. When we remember that these early laborers used only simple microscopes, generally of their own construction, we must admire their patient industry, skilful manipulation, and accurate judgment. In these respects they are models to all microscopists.

Within the last quarter of a century microscopic observers may be numbered by thousands, and some have attained an eminent reputation. At the present day, in Germany, England, France, and the United States, the most careful and elaborate investigations are being made, older observations are repeated and corrected, new discoveries are rapidly announced, and the most hidden recesses of nature are being explored.

It is proposed in this treatise to give such a résumé of microscopy as shall enable the student in any department to pursue original investigations with a general knowledge of what has been accomplished by others. To this end a comprehensive view of the necessary instruments and details of the art, or what the Germans call technology, is first given, and then a brief account of the application of the microscope to various branches of science, especially considering the needs of physicians and students of medicine.

CHAPTER II.

THE MICROSCOPE.

The Simple Microscope.—The magnifying power of a glass lens (from *lens*, a lentil; because made in the shape of its seeds) was doubtless known to the ancients, but only in modern times has it been applied in scientific research.

The forms of lenses generally used are the *double convex*, with two convex faces; *plano convex*, with one face flat and the other convex; *double concave*, with two concave faces; *plano-concave*, with one flat and one concave face; and the *meniscus*, with a concave and a convex face.

In the early part of the seventeenth century very minute lenses were used, and even small spherules of glass. Many of the great discoveries of that period were made by these means. A narrow strip of glass was softened in the flame of a spirit-lamp and drawn to a thread, on the end of which a globule was melted and placed in a thin folded plate of brass, perforated so as to admit the light. Some of these globules were so small as to magnify several hundred diameters. Of course, they were inconvenient to use, and larger lenses, ground on a proper tool, were more common.

The magnifying power of lenses depends on a few simple

optical laws, concerning refraction of light, allowing the eye to see an object under a larger visual angle; so that the power of a simple microscope is in proportion to the shortness of its focal length, or the distance from the lens to the point where a distinct image of the object is seen. This distance may be measured by directly magnifying an object with the lens, if it be a small one, or by casting an image of a distant window, candle, etc., upon a paper or wall. The focus of the lens is the point where the image is most distinct. Different persons see objects naturally at different distances, but ten inches is considered the average distance for the minimum of distinct vision. A lens, therefore, of two inches focal length, magnifies five diameters; of one inch focus, ten diameters; of one-half inch, twenty diameters; of one-eighth inch, eighty diameters; etc.

Simple microscopes are now seldom used, except as hand magnifiers, or for the minute dissection and preparation of objects. They are used for the latter purpose, when suitably mounted with a convenient arm, mirror, etc., because of the inconvenience of larger and otherwise more perfect instruments.

Single lenses, of large size, are also used for concentrating the light of a lamp on an object during dissection, or on an opaque object on the stage of a compound microscope.

There are imperfections of vision attending the use of all common lenses, arising from the spherical shape of the surface of the lens, or from the separation of the colored rays of light when passing through such a medium. These imperfections are called respectively spherical and chromatic aberration. To lessen or destroy these aberrations, various plans have been proposed by opticians. For reducing spherical aberration, Sir John Herschel proposed a doublet of two plano-convex lenses, whose focal lengths are as 2.3 to 1, with their convex sides together;

and Mr. Coddington invented a lens in the form of a sphere, cut away round the centre so as to assume the shape of an hour-glass. This latter, in a convenient setting, is one of the best pocket microscopes. Dr. Wollaston's doublet consists of two plano-convex lenses, whose focal lengths are as 1 to 3, with the plane sides of each and the smallest lens next the object. They should be

Fig. 1.

Holland's Triplet.

about the difference of their focal lengths apart, and a diaphragm or stop—an opaque screen with a hole in it—placed just behind the anterior lens. This performs admirably, yet has been further improved by Mr. Holland by making a triplet of plano-convex lenses (Fig. 1), with the stop between the upper lenses.

The Compound Microscope consists essentially of two convex lenses, placed some distance apart, so that the image made by one may be magnified by the other. These are called the object-glass and the eye-glass. In Fig. 2, A is the object-glass, which forms a magnified image at C, which is further enlarged by the eye-glass B. An additional lens, D, is usually added, to enlarge the field of view. This is called the field-glass. Its office, as in the figure, is to collect more of the rays from the object-glass and form an image at F, which is viewed by the eye-glass.

Owing to chromatic aberration, an instrument of this kind is still imperfect, presenting rings of color round the edge of the field of view as well as at the edge of the magnified image of an object, together with dimness and confusion of vision. This may be partly remedied by a small hole or stop behind the object-glass, which reduces the aperture to the central rays alone, yet it is still un-

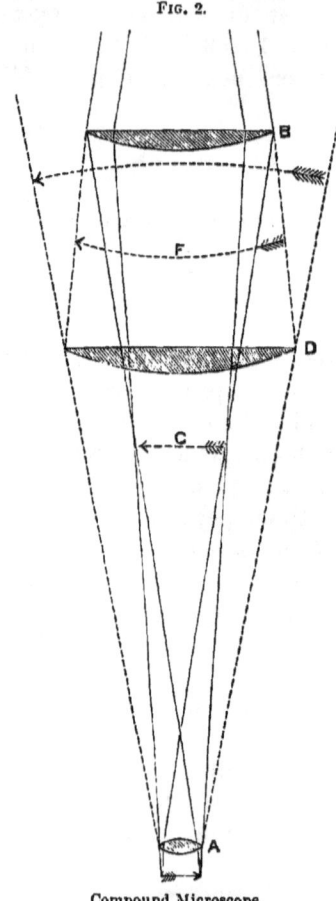

Compound Microscope.

satisfactory. Some considerable improvement may result from using Wollaston's doublet as an object-glass, but the achromatic object-glasses now supplied by good opticians leave nothing to be desired.

Object-glasses.—A general view of an achromatic object-glass is given in Fig. 3. It is a system of three pairs of lenses, 1, 2, 3, each composed of a double convex of crown glass and a plano-concave of flint. *a, b, c*, represents the angle of aperture, or the cone of rays admitted. It is unnecessary to consider the optical principles which underlie this construction. Different opticians have different formulæ and propose various arrangements of lenses, and there is room for choice among the multitude of microscopes presented for sale. For high powers, the German

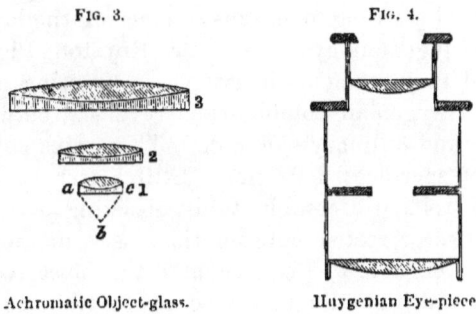

Achromatic Object-glass.   Huygenian Eye-piece.

and French opticians have lately proposed a principle of construction which is known as the immersion system. It consists in the interposition of a drop of water between the front lens of the objective and the covering glass over the object. This form of object-glass is coming into general use. For the more perfect performance of an objective, it is necessary that it should be arranged for correcting the effect of different thicknesses of covering glass. This is accomplished by a fine screw movement, which brings the front pair of lenses (1, Fig. 3) nearer or further from the object. In this way the most distinct and accurate view of an object may be obtained.

Eye-pieces.—The eye-piece usually employed is the Huygenian, or negative eye-piece (Fig. 4). This is composed of two plano-convex lenses, with their plane sides next the eye. Their focal lengths are as 1 to 3, and their distance apart half the sum of their focal distances. Several of these, having different magnifying powers, are supplied with good microscopes. It is best to use a weak eye-piece, increasing the power of the instrument by stronger objectives when necessary. Kellner's eye-piece has the lens next the eye made achromatic. The periscopic eye-piece of some of the German opticians has both lenses double convex. This gives a larger field of view with some loss of accurate definition. For high powers, I have used a strong meniscus in place of the lower lens in the Huygenian eye-piece. Dr. Royston Pigott has suggested improvements in eye-pieces by using an intermediate Huygenian combination, reversed, between the objective and ordinary eye-piece. This gains power, but somewhat sacrifices definition. Still better, he has proposed an aplanatic combination, consisting of a pair of slightly overcorrected achromatic lenses, mounted midway between a low eye-piece and the objective. This has a separating adjustment so as to traverse two or three inches. The focal length of the combination varies from one and a half to three-fourths of an inch. The future improvement of the microscope must be looked for in this direction, since opticians seem to have approached the limit of perfection in high power objectives, some of which have been made equivalent to $\frac{1}{80}$th or $\frac{1}{100}$th of an inch focal length. As an amplifier, I have used a double concave lens of an inch in diameter and a virtual focus of one and a half inches between the object-glass and the eye-piece. If the object-glass be a good one, this will permit the use of a very strong eye-piece with little loss of defining power, and greatly increase the apparent size of the object.

Mechanical Arrangements.—The German and French opticians devote their attention chiefly to the excellence of their glasses, while the mechanical part of their instruments is quite simple, not to say clumsy. They seem to proceed on the principle that as little as possible should be done by mechanism, which may be performed by the hand. It is different with English and American makers, some of whose instruments are the very perfection of mechanical skill. The disparity in cost, however, for instruments of equal optical power is quite considerable.

Certain mechanical contrivances are essential to every good instrument. The German and French stands are usually vertical, but it is an advantage to have one which can be inclined in any position from vertical to horizontal. There should be steady and accurate, coarse and fine adjustments for focussing; a large and firm stage with ledge, etc., and with traversing motions, so as to follow an object quickly, or readily bring it into the field of view; also a concave and plane mirror with universal joints, capable of being brought nearer or farther from the stage, or of being turned aside for oblique illumination. Steadiness, or freedom from vibration, is of the utmost importance in the construction, since every unequal vibration will be magnified by the optical power of the instrument.

Among so many excellent opticians it would be impossible to give a complete list of names whose workmanship is wholly reliable, yet among the foremost may be mentioned Tolles, of Boston; Wales, of Fort Lee, N. J.; Grunow, of New York; and Zentmayer, of Philadelphia; Powell & Leland, Ross and Smith, Beck & Beck, of London; Hartnack and Nachet, of Paris; Merz, of Munich; and Gundlach, of Berlin. The optical performance of lenses from these establishments is first class, and the mechanical work of their various models good. The finest instruments from these makers, with complete appliances, are quite costly, except the Germans and French, whose arrangements, as we have said, are more simple. Cheaper instruments, however, are made by English and American opticians, some of which are very fine.

Opticians divide microscopes into various classes, according to the perfection of their workmanship or the accessories supplied. The best first-class instruments have

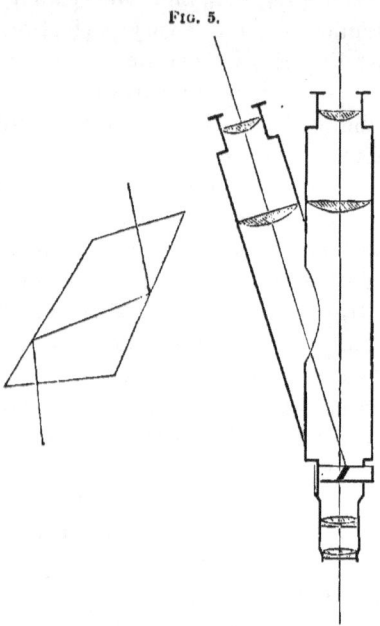

Fig. 5.

Wenham's Prism for the Binocular Microscope.

a great variety of objectives and eye-glasses, mechanical stage with rack-work; a sub-stage with rack for carrying various illuminators: a stand of most solid construction; and every variety of apparatus to suit the want or wish of the observer. They are great luxuries, although not essential to perfect microscopic work. The second class, or students' microscopes, have less expensive stands, but equal optical powers, with first-class instruments. The

FIG. 6.

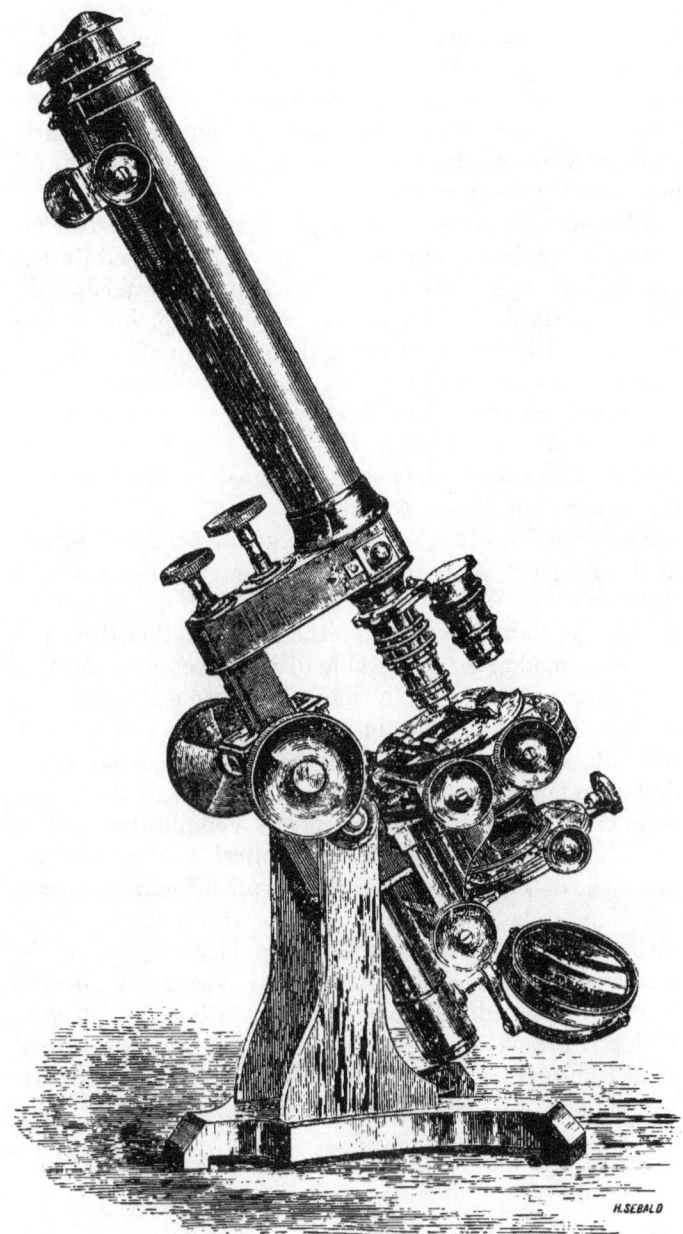

Collins's Harley Binocular Microscope.

third or fourth classes of instruments are intended for popular and educational use, and are fitted not only with stands of more simple workmanship, but with cheaper lenses, although often very good. Some French achromatic objectives, adapted to this class, are suitable for all but the very finest work.

Binocular Microscopes.—The principle of the stereoscope has been applied to the microscope, so as to permit the use of both eyes. The use of such an instrument with low or medium powers is very satisfactory, but is less available with objectives stronger than one-half inch focus. There are two ways of accomplishing a stereoscopic effect in the microscope. The first and most common is by means of Wenham's prism (Fig. 5), placed above the objective, and made to slide so as to transform the binocular into a monocular microscope.

The second mode is to place an arrangement of prisms in the eye-piece, so as to refract one-half the image to the right and the other half to the left, which are viewed by the corresponding eyes. In either construction there is a provision made for the variable distance between the eyes of different observers. In the frontispiece is a representation of Zentmayer's grand American microscope, which will afford a good idea of the external appearance of a first-class binocular microscope. Students' and third-class microscopes, as before said, are less complicated and of more moderate cost. The mechanical and optical performance of Zentmayer's large instrument leaves scarcely anything to be desired. Instead of the more expensive rack-work stage, a simple form, originally invented by Dr. Keen, of Philadelphia, and copied by Nachet and others, is often employed. It consists of a rotating glass disk, to which is attached a spring, or a V-shaped pair of springs, armed with ivory knobs, which press upon a glass plate in the object-carrier. The motion is exceedingly smooth and effective.

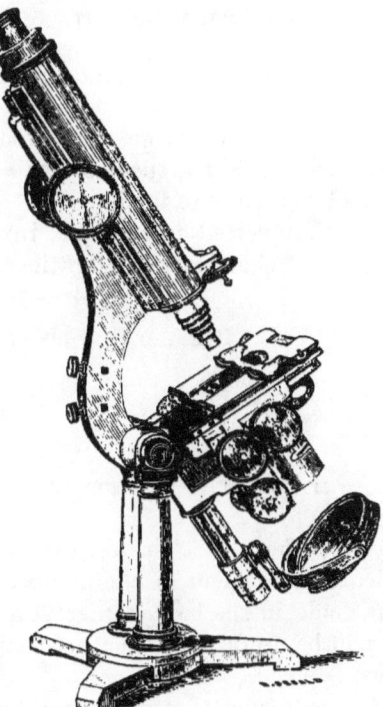

Beck's Large Compound Microscope.

FIG. 8. FIG. 9.

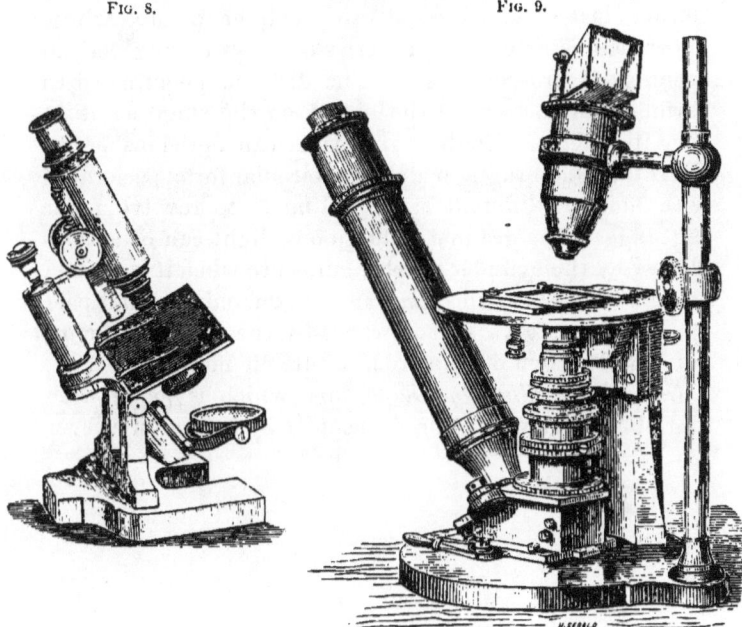

Hartnack's Small Model Microscope.   Nachet's Inverted Microscope.

Fig. 6 shows Collins's Harley binocular microscope, a good second class instrument.

Fig. 7 represents Beck's large compound miscroscope (monocular); and Fig. 8, Hartnack's small model microscope, with the body made to incline.

Fig. 9, Nachet's inverted microscope, invented by Dr. Lawrence Smith for chemical investigations.

CHAPTER III.

MICROSCOPIC ACCESSORIES.

In addition to the object-glasses, eye-glasses, mirror, and mechanical arrangement of the microscope, to which reference was made in the last chapter, several accessory instruments will be useful and even necessary for certain investigations.

The Diaphragm, for cutting off extraneous light when viewing transparent objects, is generally needed. In some German instruments it consists of a cylinder or tube, whose upper end is fitted with a series of disks having central openings of different sizes. The disk can be adjusted to variable distances from the object on the stage so as to vary its effects. English and American opticians prefer the rotary diaphragm, which is of circular form, perforated with holes of different sizes, and made to revolve under the stage. The gradual reduction of light can be accomplished by the cylinder diaphragm, since when it is pushed up so as to be near the stage it cuts off only a small part of the cone of rays sent upwards by the concave mirror, but, when drawn downwards, it cuts off more.

Collins's Graduating Diaphragm, which is made with four shutters, moving simultaneously by acting on a lever handle, so as to narrow the aperture, accomplishes the end most perfectly. (Fig 10.)

Fig. 10.

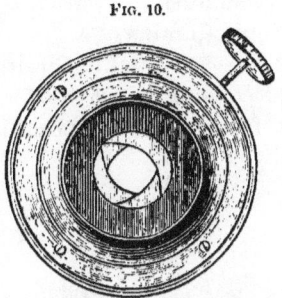

Collins's New Graduating Diaphragm.

Beck's Iris Diaphragm is a further improvement of this sort.

Condensers.—The loss of light resulting from the employment of high powers has led to several plans for condensing light upon the object. Sometimes a plano-convex lens, or combination of lenses, is made to slide up and down under the stage. A *Kellner's eye piece*, or some

Fig. 11.

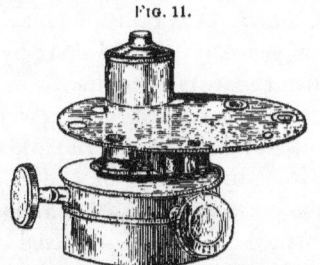

Smith and Beck's Achromatic Condenser.

similar arrangement, especially if fitted with a special diaphragm, containing slits and holes, some of the latter having central stops, is of very great use. First-class instruments are fitted up with *achromatic condensers* (Fig. 11), carrying revolving diaphragms, some of whose apertures are more or less occupied by stops, or solid disks, so as to leave but a ring of space for light to pass through. The effect of these annular diaphragms is similar to an apparatus for oblique illumination.

The *Webster condenser* is similar in its optical parts to the Kellner eye-piece, and is provided with a diaphragm plate, with stops for oblique illumination, as well as a

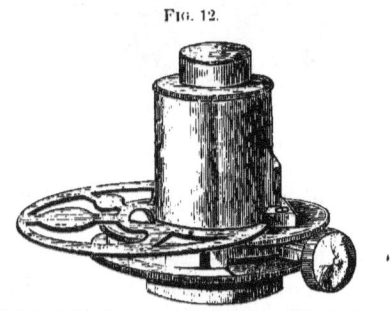

Fig. 12.

Webster's Condenser, with Graduating Diaphragms.

graduating diaphragm for the regulation of the central aperture. This is a most useful accessory. (Fig. 12.)

Oblique Illuminators.—Certain fine markings on transparent objects can scarcely be made out by central illumination, but require the rays to come from one side, so as to throw a shadow. Sometimes this is well accomplished by turning the mirror aside from the axis of the microscope, and sometimes by the use of one of the condensers referred to above. *Amici's prism*, which has both plane and lenticular surfaces, is sometimes used on one side and under the stage, in lieu of the mirror. For obtaining very oblique pencils of light the *double hemispherical condenser* of Mr. Reade has been invented. It is a hemispherical lens of about one and a half inch diameter, with its flat side next the object, surmounted by a smaller lens of the same form, the flat side of which is covered with a thin diaphragm, having an aperture or apertures close to its margin. These apertures may be V-shaped, extending to about a quarter of an inch from the centre.

If the microscope has a mechanical stage, with rack-work, or is otherwise too thick to permit the mirror to be turned aside for very oblique illumination, *Nachet's prism* will prove of service. I have also contrived a useful oblique illuminator for this purpose, by cementing with Dammar varnish a plano-convex lens on one face of a totally-reflecting prism, and near the upper edge of the other side (at 90°) an achromatic lens from a French triplet. The prism is made to turn on a hinge, so that an accurate pencil of light may fall on the object at any angle desired.

Dark-ground Illuminators.—Some beautiful effects are produced, and the demonstration of some structures aided, by preventing the light condensed upon the object from entering the object-glass. In this way the object appears

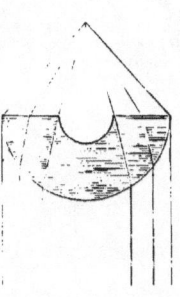

Fig. 13.

Nobert's Illuminator.

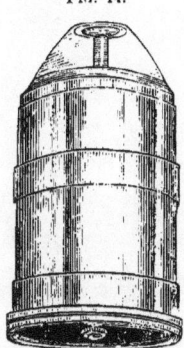

Fig. 14.

Parabolic Illuminator.

self-luminous on a black ground. For low powers this can be easily done by turning aside the concave mirror as in oblique illumination, or by employing *Nobert's illuminator*, which is a thick plano-convex lens, in the convex surface of which a deep concavity is made. The plane side is next the object. This throws an oblique light all round the object. A substitute for this, called a *spot lens*, is often used, and differs only from Nobert's in having a central black stop on the plane side instead of a concavity (Fig. 13). A still greater degree of obliquity suitable for high powers must be sought by the use of the *parabolic illuminator* (Fig. 14). This is usually a paraboloid of glass, which reflects to a focus the rays which fall upon its internal surface, while the central rays are stopped.

Illuminators for Opaque Objects.—Ordinary daylight is hardly sufficient for the illumination of opaque objects,

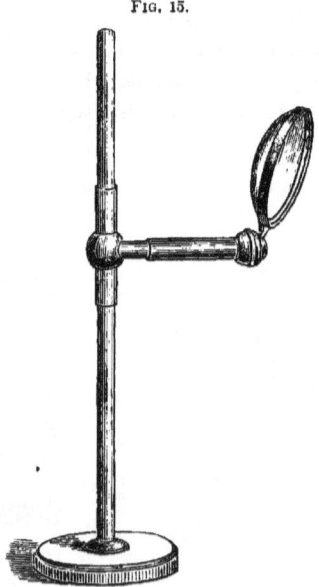

Fig. 15.

Bull's-eye Condenser.

so that microscopists resort to concentrated lamplight, etc. Gas, paraffine, and camphene lamps, have been variously modified for this purpose, but few are better than the German student's Argand lamp for petroleum or kerosene oil, as it is called. To concentrate the light from such a source a *condensing lens* is used, either attached to the microscope or mounted on a separate stand. Sometimes a *bull's-eye condenser* is used for more effective illumination (Fig. 15). This is a large plano-convex lens of short focus, mounted on a stand. For such a lens the position of least spherical aberration is when its convex side is towards parallel rays; hence, in daylight, the plane side should be next the object. But, if it is desired to render the diverg-

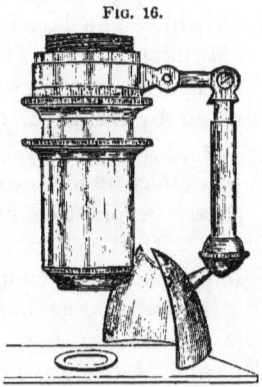

Fig. 16.

Parabolic Speculum.

ing rays of a lamp parallel, the plane side should be next the lamp, and rather close to it. The use of this condenser will also commend itself, when used as last referred to, in microscopic dissection. It will throw a bright light from the lamp directly on the trough, watch-glass, etc., in which the specimen is being prepared. The *Lieberkuhn*, or a concave speculum attached to the object-glass, and reflecting the light from the mirror directly upon the object, is one of the oldest contrivances for the illumination of opaque objects; but the most convenient instrument is the *parabolic speculum* (Fig. 16), a side mirror with a parabolic surface attached to the objective. For high powers, a lateral aperture above the objective has been made to throw the light down through the object-glass itself by means of a small reflector, as devised by Prof. Smith, or a disk of thin glass, as in *Beck's vertical illuminator*. This latter is attached to an adapter interposed between the objective and the body of the microscope.

Instruments for Measuring and Drawing Objects.—Screw micrometers are sometimes used with the microscope, as with the telescope, for the measurement of objects; but the less expensive and simpler glass micrometers have generally superseded them. The latter are of two sorts, the stage and the ocular micrometer. The *stage micrometer* is simply a glass slide, containing fine subdivisions of the inch, line, etc., engraved by means of a diamond point. In case the rulings are $\frac{1}{100}$ths and $\frac{1}{000}$ths of an inch, it is evident that an object may be measured by comparison with the divisions; yet, in practice, it is found inconvenient to use an object with the stage micrometer in this way, and it will be found better to combine its use with that of the drawing apparatus, as hereafter described. The ocular, or *eye-piece micrometer*, is a ruled slip of glass in the eye-piece. Its value is a relative one, depending on the power of the objective and the length of the microscope tube. By comparing the divisions with those of the stage micrometer their value can be readily ascertained. Thus, if five spaces of the eye-piece micrometer cover one space of the stage micrometer, measuring $\frac{1}{1000}$th of an inch, their value will be $\frac{1}{500}$th of an inch each.

Different standards of measurement are used in different countries. English and American microscopists use the inch. In France, and generally in Germany, the Paris line or the millimetre is used. The millimetre is 0.4433 of a Paris line and 0.4724 of an English line ($\frac{1}{12}$th of an inch).

In the French system the fundamental unit is the metre,

which is the ten-millionth part of the quadrant of the meridian of Paris. The multiples are made by prefixing Greek names of numbers, and the subdivisions by prefixing Latin names. Thus, for decimal multiples, we have *deco, hecto, kilo, and myrio;* and, for decimal subdivisions, *deci, centi,* and *milli.* The following may serve for converting subdivisions of the metre into English equivalents:

    A millimetre equals 0.03937 English inches.
    A centimetre   "   0.39371   "
    A decimetre   "   3.93708   "
    One inch = 2.539954 centimetres, or 25.39954 millimetres.

For drawing microscopic objects the *camera lucida* will be found useful. This is a small glass prism attached to the eye-piece. The microscope is inclined horizontally,

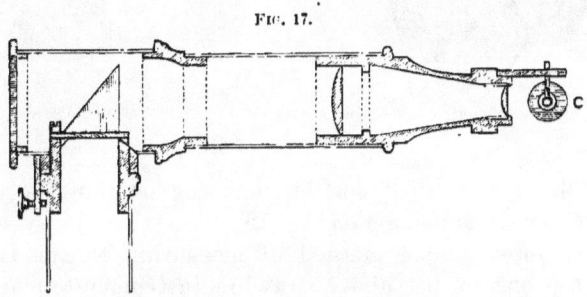

Fig. 17.

Oberhauser's Drawing Apparatus.

and the observer, looking into the prism, sees the object directly under his eye, so that its outlines may be drawn on a piece of paper placed on the table. Some practice, however, is needed for satisfactory results. For the upright stands of German and French microscopes, the camera lucida of Chevalier & Oberhauser is available. This is a prism in a rectangular tube, in front of which is the eye-piece, carrying a small glass prism (c, Fig. 17), surrounded by a black metal ring. A paper placed beneath is visible through the opening in the ring, and the image reflected by the prism upon it can be traced by a pencil. It is necessary to regulate the light so that the point of the pencil may be seen.

Dr. Beale has recommended, in lieu of the camera lucida, a piece of slightly tinted plate glass (Fig. 18), placed in a short tube over the eye-piece at an angle of 45°. This is a cheap and effective plan. A similar purpose is served

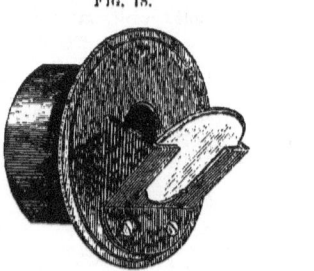

Beale's Tint-glass Camera.

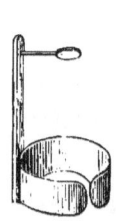

Sœmmering's Steel Disk.

by a little steel disk, smaller than the pupil of the eye, placed at the same angle (Fig. 19).

The most simple method of measuring objects is to employ one of the above drawing instruments, placing first on the microscope stage an ordinary micrometer, and tracing its lines on the paper. Then the outline of the object can be traced and compared with the lines. The magnifying power of an object-glass can also be readily found by throwing the image of the lines in a stage micrometer upon a rule held ten inches below the eye-piece, looking at the magnified image with one eye and at the rule with the other. Dr. Beale strongly urges observers to delineate their own work on wood or stone, since they can do it more exactly and truthfully than the most skilled artists who are unfamiliar with microscopic manipulation.

Other accessory apparatus, such as a *frog-plate*, for more readily observing the circulation in a frog's foot; an *animalcule cage*, or live box; a *compressorium*, for applying pressure to an object; fishing tubes; watch-glasses; growing-slides, etc., will commend themselves on personal inspection.

For preventing the evaporation of fluids during observation, Recklinghausen invented the *moist chamber* (Fig. 20), consisting of a glass ring on a slide, to which is fastened a tube of thin rubber, the upper end of which is fastened round the microscope tube with a rubber band.

Recklinghausen's Moist Chamber.

A simpler form of moist chamber may be made by a glass ring cemented on a slide. A few drops of water cautiously put on the inner edge of the ring with a brush, or a little moist blotting-paper may be placed inside. The object (as a drop of frog's blood, etc.) may then be put on a circular thin cover, which is placed inverted on the ring. A small drop of oil round the edge of the cover keeps it air and water-tight.

Somewhat similar to the above is Stricker's *gas chamber* (Fig. 21). On the object-slide is a ring of glass, or putty, with its thin cover. Through this ring two glass tubes are cemented, one of which is connected with a rubber tube for the entrance of gas, while the other serves for its exit.

For the study of phenomena in the fluids, etc., of warm-blooded animals, we need, in addition to the moist chamber, some way of keeping the object warm. This may be roughly done by a perforated tin or brass plate on the stage, one end of which is warmed by a spirit-lamp. A piece of cocoa butter or wax will show by its melting when the heat is sufficient. Schultze's *warm stage* is a more satisfactory and scientific instrument. It is a brass plate to fit on the stage, perforated for illumination, and connected with a spirit-lamp and thermometer, so that

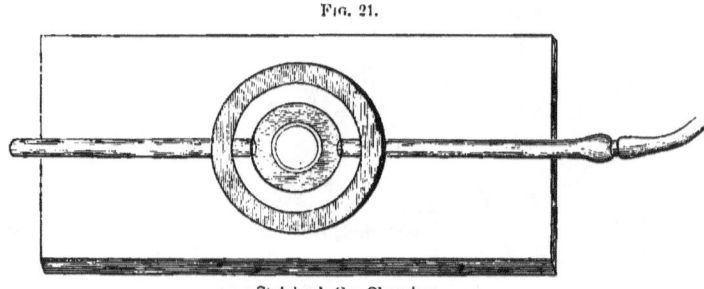

Fig. 21.

Stricker's Gas Chamber.

the amount of heat may be exactly regulated. Other arrangements have been proposed to admit a current of warm water, or for the passage of electricity through an object while under observation, which are scarcely necessary to describe.

The Polariscope.—The nature and properties of polarized light belong rather to a treatise on optics or natural philosophy than to a work like the present, yet a very brief account may not be out of place. We premise, then, that every ray or beam of common light is supposed to have at least two sets of vibrations, vertical and horizontal. As these vibrations have different properties, the ray when divided is said to be *polarized*, from a fancied resemblance to the poles of a magnet. The division of the vibrations may be effected (*i. e.*, the light may be polarized) in various ways. For the microscope the *polarizer* is a Nichol's prism, composed of a crystal of Iceland spar, which has been divided and again cemented with Canada balsam, so as to throw one of the doubly refracted rays aside from the field of view (Fig. 22). Such a prism is mounted in a short tube and attached to the under side of the stage. In order to distinguish the effects of polarized light, an *analyzer* is also needed. This usually consists of another

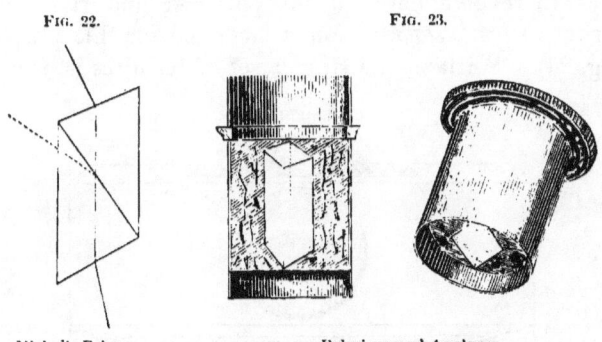

Fig. 22. Fig. 23.

Nichol's Prism. Polarizer and Analyzer.

similar Nichol's prism, attached either to the eye-piece or just above the objective. The latter position gives a larger field, but the former better definition. Fig. 23 shows the polarizer and the analyzer. The polarizer is improved by the addition of a convex lens next the object. Hartnack has also improved the eye-piece analyzer by adding a graduated disk and vernier.

When the polarizer and analyzer have been put in place, they should be rotated until their polarizing planes are parallel, and the mirror adjusted so as to give the most intense light. If now the polarizing planes are placed at right angles, by turning one of them 90°, the field is rendered dark, and doubly refracting bodies on the stage of the microscope appear either illuminated or in colors. If a polarized ray passes through a doubly refracting film, as of selenite, it forms two distinct rays, the ordinary and the extraordinary ray. Each of these will be of different colors, according to the thickness of the film. If one be red, the other will be green, these colors being complementary. By using the analyzer one of these rays is alternately suppressed, so that on revolving the apparatus the green and red rays appear to alternate at each quarter of a circle. Films of selenite are often mounted so as to revolve between the polarizer and the stage. Darker's *selenite stage* is sometimes used for this purpose (Fig. 24). With such a stage a set of selenites is usually

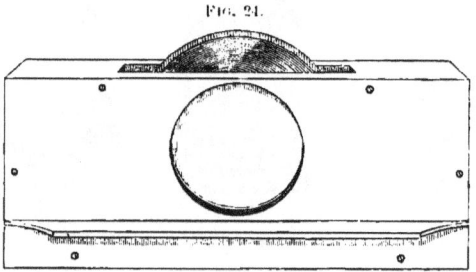

Fig. 24.

Darker's Selenite Stage.

supplied, giving the blue, purple, and red, with their complementary colors, orange, yellow, and green. By this combination all the colors of the spectrum may be obtained. The selenite disks generally have engraved on them the amount of retardation of the undulations of white light, thus: $\frac{1}{4}$, $\frac{3}{4}$, and $\frac{2}{3}$. If these are placed so that their positive axes (marked PA) coincide, they give the sum of their combined retardations.

The Microspectroscope.—Ordinary spectrum analysis, by determining the number and position of certain narrow lines in the spectra of luminous bodies, called Fraunhofer's lines, enables the chemist to identify different substances. The object of the microspectroscope is different. It enables us to distinguish substances by the absence of certain rays in the spectrum, or, in other words, to judge of substances by a scientific examination of their color. The color of a body seen with the naked eye is the general impression made by the transmitted light, and this may be the same although the compound rays may differ

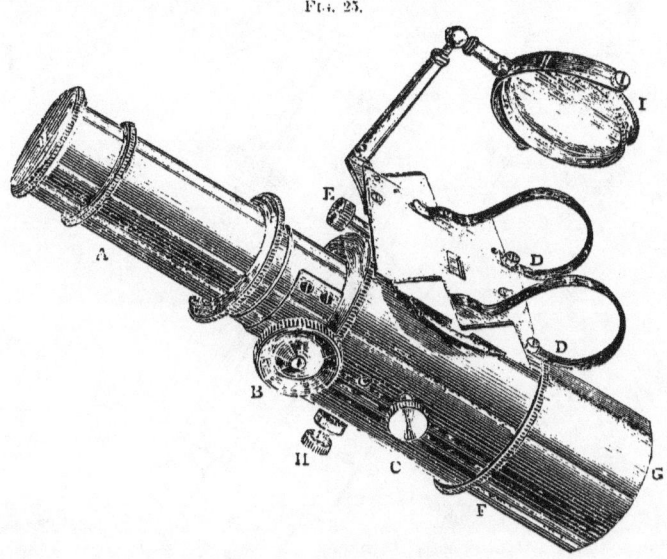

Fig. 25.

The Sorby-Browning Microspectroscope.

greatly, so that colors which seem absolutely alike may be distinguished by their spectra. Many solutions are seen to absorb different colors in very definite parts of the spectrum, forming absorption bands or lines, varying in width and intensity according to the strength of the solution. The instrument usually employed consists of a direct-vision spectrum apparatus attached to the eye piece of the microscope, which shows the principal Fraunhofer lines by daylight, or a spectrum of the light transmitted by any object in the field of view. A reflecting prism is placed under one-half of the slit of the apparatus so as to transmit from a side aperture a standard spectrum for comparison. In Fig. 25, A is a brass tube carrying the compound direct-vision system of five prisms and an achromatic lens. This tube is moved by the milled head

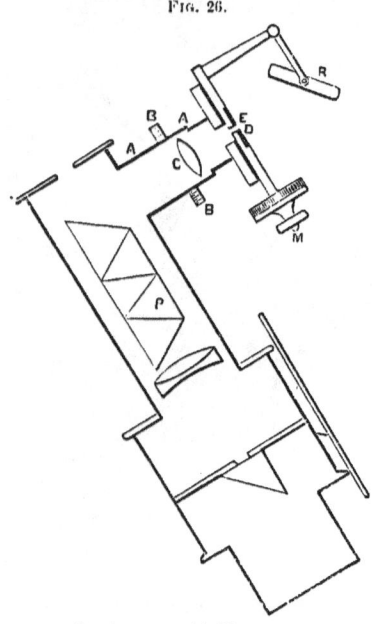

FIG. 26.

Spectroscope with Micrometer.

B, so as to bring to a focus the different parts of the spectrum. This is important when the bands or lines to be examined are delicate. D is the stage on which objects for comparison are placed. The light passing through them from the mirror I, goes through a side opening to a reflecting prism which covers a part of a slit in the bottom of the tube A. This slit is opened and shut by means of the screws C and H. Fig. 26 shows the internal arrangement of the prisms and lens, together with a micrometer for measuring the position of lines or absorption bands. To use the microspectroscope, remove the tube A, with the prisms, and insert the tube G in the place of the eye-piece of the microscope. With the lowest power object-glass which is suitable, and the slit opened wide by the screw H, the object on the stage of the microscope, illuminated by the mirror or condenser, is brought to a focus, the tube A replaced and adjusted for focus by the screw B, while the slit is regulated by C and H until a well-defined spectrum is seen. To determine the position of the absorption lines, remove the upper cover of the tube A and replace it with that carrying the micrometer represented in Fig. 26. The mirror illuminates a transparent line or cross, whose image is refracted by a lens C, movable by a screw B, and reflected at an angle of 45° from the upper surface of the prisms, so as to be seen upon the spectrum. By means of the micrometer screw M, this is made to move across the spectrum, so that the distance between the lines may be determined. In order to compare the results given by different instruments, the observer should measure the position of the principal Fraunhofer lines in bright daylight, and mark them on a cardboard scale, which may be preserved for reference. By comparing the micrometric measurement of lines in the spectrum of any substance observed by artificial light with such a scale, their position may readily be seen.

In using the microspectroscope some objects require a diaphragm of small size, and others, especially with the $1\frac{1}{2}$ or 2-inch objective, a cap with a hole $\frac{1}{6}$th of an inch in diameter over the end of the microscope, to prevent extraneous light from passing through the tube.

Nose-piece.—For the purpose of facilitating observations with objectives of different powers a revolving nose-piece has been contrived, carrying two, three, or four objectives, which may be brought quickly into the axis of the instrument.

Object-finders.—It is sometimes tedious to find a small object on a slide, particularly with high powers, and a number of contrivances, as Maltwood's finder, have been proposed for this end. A very simple method, however, may serve. Mark on the stage two crosses, one like the sign of addition +, and the other like the sign of multiplication ×, and, when the object is found, mark the slide to correspond with the marks below. If the stage be a mechanical one it will be necessary to arrange it in the previous position.

Microscopic Photography.—Many European experimenters have succeeded in taking microscopic photographs, but a great advance in this direction has been made under the direction of the medical department of the United States army at Washington. Lieutenant-Colonel Woodward has succeeded in furnishing permanent records of many details of structure, which exhibit the very perfection of art. In a work like the present a full account of the apparatus and methods employed would be out of place. Dr. Beale's *How to Work with the Microscope*, and the reports issued from the Surgeon-General's office at Washington, will give the details.

CHAPTER IV.

USE OF THE MICROSCOPE.

Care of the Instrument.—But little satisfaction will be secured in microscopic work for any length of time without scrupulous care of the lenses, etc., belonging to the instrument, and habits of this kind should be early acquired. When in frequent use the microscope should be seldom packed away in its case, as a certain necessary stiffness of motion in its various parts might thereby be lessened. Yet it should be kept free from dust and damp. A bell-glass cover, or glass case, or a cabinet which will admit the reception of the instrument in a form ready for immediate use, is desirable. Before using, the condition of objective and eye-piece should be examined as well as of the mirror, and dust or dampness removed. Another examination should be made before the microscope is put away.

Stains on the brass-work may be removed by a linen rag, and dust on the mirror and lenses by a fine camel's-hair brush, or very soft and clean chamois skin. Frequent wiping will injure the polish of the lenses.

The upper surfaces of the lenses in the eye-pieces and the mirror will need the most frequent attention The objectives, if carefully handled and kept in their boxes when not in use, will seldom require cleaning. If the front of the objective becomes accidentally wet with fluid it should be at once removed, and, when reagents are used, great care should be taken to prevent contact with the front of the lens.

Care of the Eyes.—Continuous observation, especially by lamplight, and with high powers, has doubtless a tendency to injure the sight. To cease work as soon as fatigue begins is, however, a simple but certain rule for protection. This time will vary greatly, according to the general tone and vigor of the observer. It is also important to use the eyes alternately if a monocular instrument is employed, as otherwise great difference both in the focus and in the sensitiveness of the eyes will result. The habit of keeping the unemployed eye open is a good one, and, though troublesome at first, is not difficult to acquire. It is well to protect the eye from all extraneous light, and to exclude every part of the object except that which is under immediate observation. The diaphragm will serve this end as well as modify the quality of the light. For very delicate observations a dark shade over the stage, which may be fastened by an elastic ring to the microscope-tube, so as to shut off extraneous light, will be useful.

Table, etc.—The microscopist's work-table should be large and massive, so as to be convenient and free from vibration. Drawers for accessories and materials used in preparing and mounting objects are also desirable, as well as a few bell-glasses for secluding objects from dust. Reagents should always be removed from the table after use and kept in another place.

Light.—Dr. Carpenter has well said, "Good daylight is to be preferred to any other kind of light, but good lamp-light is preferable to bad daylight." A clear blue sky gives light enough for low powers, but a dull white cloudiness is better. The direct rays of the sun are too strong, and should be modified by a white curtain, reflection from a surface of plaster of Paris, or, still better, by passing through a glass cell containing a solution of ammonio-sulphate of copper.

Various kinds of lamps have been contrived for microscopic use; among the best are the German and French "student's reading lamps," which burn coal oil or petroleum. It is often useful to moderate such a light by the use of a chimney of blue glass, or by a screen of blue glass between the flame and the object. Dr. Curtis contrived a useful apparatus, consisting of a short petroleum lamp placed in an upright, oblong box. On one side of the box is an opening occupied with blue glass; on another side the opening has ground-glass, as well as a piece colored blue, and a plano-convex lens so placed as to condense the light thus softened to a suitable place on the table.

As a general rule the light should come from the left side, and that position assumed or inclination given to the instrument which is most comfortable to the observer.

English and American microscopists prefer an inclined microscope, while the German and French instruments being usually vertical do not permit this arrangement.

Adjustment.—The details of microscopic adjustments are only to be learned by practice, yet a few directions may be instructive. The selection of the objectives and eye-pieces depends on the character of the object. As a general rule, the lowest powers which will exhibit an object are the best. It is best to use weak eye-pieces with the stronger objectives, yet much depends on the perfection of the glasses employed.

The focal adjustment can be made with the coarse adjustment or quick motion when low powers are employed; but for higher powers the fine adjustment screw is essential. Care must be taken not to bring the objective into close or sudden contact with the thin glass cover over the object, and, in changing object-glasses, the microscope body should be raised from the stage by the coarse adjustment.

The actual distance between the object and object-glass is much less than the nominal focal length, so that the 1 inch objective has a working distance of about $\frac{1}{2}$ an inch, the $\frac{1}{8}$th of about $\frac{1}{40}$th of an inch, while shorter objectives require the object to be covered with the thinnest glass.

Sometimes, in high powers, and especially with immersion-lenses, an adjustment of the object-glass is necessary in order to suit the thickness of the glass cover. With thick covers the individual lenses must be brought nearer to each other, and, with very thin covers, moved farther apart.

If immersion-objectives be employed a drop of water is placed on the glass cover with a glass rod or camel's-hair pencil, and a second drop on the lens. The lens and object are then approximated till the drops flow together and the focus is adjusted. By turning the screw of the objective and using the fine adjustment the best position will be shown by the sharper and more delicate image of the object.

For other details respecting adjustment the reader is referred to the chapter on Microscopic Accessories.

Errors of Interpretation.—True science is hindered most of all by speculation and false philosophy, which often assume its garb and name, but it is also retarded by imperfect or false observation. It is much less easy to see than beginners imagine, and still less easy to know what we see. The latter sometimes requires an intellect of surpassing endowments. The sources of error are numerous, but some require special caution, and to these we now refer.

The nature of microscopic images causes error from imperfect focal adjustment. We see distinctly only that stratum of an object which lies directly in focus, and it is seldom that all parts of an object can be in focus together. Hence we only recognize at once the outline of an object, but not its thickness, and, as the parts which are out of focus are indistinct, we may readily fall into error. Glasses vary much in this respect. Some have considerable penetrating and defining power even with moderate angular aperture, and are better for general work than those more perfect instruments which give paler images and only reveal their excellencies to the practiced microscopist.

Sometimes the focal adjustment leads to error on account of the reversal of the lights and shadows at different distances. Thus the centres of the biconcave blood-disks appear dark when in focus, and bright when a little within the focus; while the hexagonal elevations of a diatom, as the *Pleurosigma angulatum*, are light when in focus, with dark partitions, and dark when just beyond the focus. From this we gather a means of discrimination, since a convex body appears lighter by raising the microscope, and a concave by lowering it.

The refractive power of the object, or of the medium in which it lies, is sometimes a source of error. Thus a human hair was long thought to be tubular, because of the convergence of the rays of light on its cylindrical convexity. A glass cylinder in balsam appears like a flat band, because of the nearly equal refractive powers of object and medium. The lacunæ and canaliculæ of bone were long considered solid, because of the dark appearance presented on account of the divergence of the rays passing through them. Their penetration with Canada balsam, however, proves them to be cavities. Air-bubbles, from refraction, present dark rings, and, if present in a specimen, seldom fail to attract the first attention of an inexperienced observer. The difference between oil-globules in water and water in oil, or air-bubbles, should be early learned, as in some organized structures oil-particles and vacuoles (or void spaces) are often interspersed. A globule of oil in water becomes darker as the object-glass is depressed, and lighter when raised; while the reverse is the case with water in oil, since the difference of refraction causes the oil particles to act as convex lenses, and those of water like concave lenses.

Other errors arise from the phenomena of motion visible under the microscope. A dry filament of cotton, or other fabric absorbing moisture, will often oscillate and twist in a curious way.

If alcohol and water are mixed, the particles suspended acquire a rapid motion from the currents set up, which continues till the fluids are thoroughly blended. Nearly all substances in a state of minute division exhibit, when suspended in fluid, a movement called the "Brownonian motion," from Dr. Robert Brown, who first investigated it. It is a peculiar, uninterrupted, dancing movement, the cause of which is still unexplained. These movements, as all others, appear more energetic when greatly magnified by strong objectives. It requires care to discriminate between such motions and the vital or voluntary motions of organized bodies.

The inflection or diffraction of light is another source of error, since the sharpness of outline in an object is thus impaired. The shadow of an opaque object in a divergent pencil of light presents, not sharp, well-defined edges, but a gradual shading off, from which it is inferred that the rays do not pass from the edge of the object in the same line as they come to it. This is in consequence of the undulatory nature of light. When any system of waves meets with an obstacle, subsidiary systems of waves will be formed round the edge of the obstacle and be propagated simultaneously with the original undulations. For a certain space around the lines in which the rays, grazing the edge of the opaque body, would have proceeded, the two systems of undulation will intersect and produce the phenomena of interference. If the opaque body be very small, and the distance from the luminous point proportionally large, the two pencils formed by inflection will intersect, and all the phenomena of interference will become evident. Thus, if the light be homogeneous, a bright line of light will be formed under the centre of the opaque object, outside of which will be dark lines, and then bright and dark lines alternately. If the light be compound solar light, a series of colored fringes will be formed. In addition to the results of inflection, oblique illumination at certain angles produces a double image, or a kind of overlying shadow, sometimes called the "diffraction spectrum," although due to a different cause. No rules can be given for avoiding errors from these optical appearances, but practice will enable one to overcome them, as it were, instinctively.

Testing the Microscope.—The defining power of an instrument depends on the correction of its spherical and chromatic aberrations, and excellence may often be obtained with objectives having but a moderate angle of aperture. It may be known by the sharp outline given to the image of an object, which is not much impaired by the use of stronger eye pieces.

Resolving power is the capability an instrument has of bringing out the fine details of a structure, and depends mainly on the angle of aperture of the objective, or the angle formed by the focus and the extremities of the diameter of the lens. On this account the increase of the angle of aperture has been a chief aim with practical opticians.

Penetrating power is the degree of distinctness with which the parts of an object lying a little out of focus may be seen. Objectives which have a large angle of aperture, and in consequence great resolving power, are often defective in penetration, their very perfection only permitting accurate vision of what is actually in focus. Hence for general purposes a moderate degree of angular aperture is desirable.

Flatness of field of view is also a necessity for accurate observation. Many inferior microscopes hide their imperfection in this respect by a contracted aperture in the eye-piece, by which, of course, only a part of the rays transmitted by the objective are available.

Object-glasses whose focal length is greater than half an inch are called low powers. Medium powers range from one-half to one-fifth of an inch focal length, and all objectives less than one-fifth are considered high powers.

For definition with low power objectives, the pollen grains of hollyhock, or the tongue of a fly, or a specimen of injected animal tissue, will be a sufficient test. The aperture should be enough to give a bright image, and the definition sufficient for a clear image. A section of wood, or of an echinus spine, will test the flatness of the field.

Medium powers are seldom used with opaque objects unless they are very small, but are most useful with properly prepared transparent objects. A good half-inch objective should show the transverse markings between the longitudinal ribs on the scales of the *Hipparchia janira*, butterfly (Plate I, Fig. 27), and the one-fourth or one-fifth should exhibit markings like exclamation points on the smaller scales of *Podura plumbea* (Plate I, Fig. 28) or *Lepidocyrtis*.

High power objectives are chiefly used for the most delicate and refined investigations of structure, and are not so suitable for general work. It is with these glasses that angular aperture is so necessary to bring out striæ, and dots, and other delicate structures, under oblique illumination. For these glasses, the best tests are the siliceous envelopes of diatoms, as the *Pleurosigma angulatum*, *Surirella gemma*, *Grammatophora subtilissima;* or the wonderful plates of glass artificially ruled by M. Nobert, and known as Nobert's test.

The latter test is a series of lines in bands, the distance between the lines decreasing in each band, until their existence becomes a matter of faith rather than of sight, since no glass has ever revealed the most difficult of them. The test plate has nineteen bands, and their lines are ruled at the following distances: Band 1, $\frac{1}{1000}$th of a Paris line (to an English inch as .088 to 1.000, or as 11 to 125). Band 2, $\frac{1}{1500}$th. Band 3, $\frac{1}{2000}$th. Band 5, $\frac{1}{3000}$th. Band 9, $\frac{1}{5000}$th. Band 13, $\frac{1}{7000}$th. Band 17, $\frac{1}{9000}$th. Band 19, $\frac{1}{10000}$th.

It is said that Hartnack's immersion system No. 10 and oblique light has resolved the lines in the 15th band, in which the distance of lines is about $\frac{1}{91000}$th of an inch.

The surface markings of minute diatoms are also excessively fine. Those of *Pleurosigma formosum* being from 20 to 32 in $\frac{1}{1000}$th of an inch; of *P. hippocampus* and *P. attenuatum* about 40; *P. angulatum* 46 to 52; *Navicula rhomboides* 60 to 111; and *Amphipleura pellucida* 120 to 130. This latter has been variously estimated at 100,000

PLATE I.

Fig. 27.

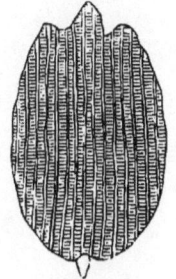

Scale of *Hipparchia Janira*.

Fig. 28.

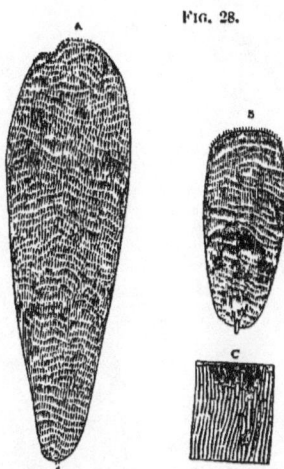

Scales of *Podura plumbea*:—A, large strongly marked scale; B, small scale more faintly marked; C, portion of an injured scale, showing the nature of the markings.

Fig. 29.

Pleurosigma angulatum:—A, entire frustule, as seen under a power of 500 diam.; B, hexagonal areolation, as seen under a power of 1300 diam.; C, the same, as seen under a power of 15,000 diam.

to 130,000 in an inch. It has been resolved by Dr. Woodward with the $\tfrac{1}{15}$th immersion of Powell and Lealand, using oblique sunlight through a solution of ammonio-sulphate of copper.

The longitudinal lines (between the transverse) of the

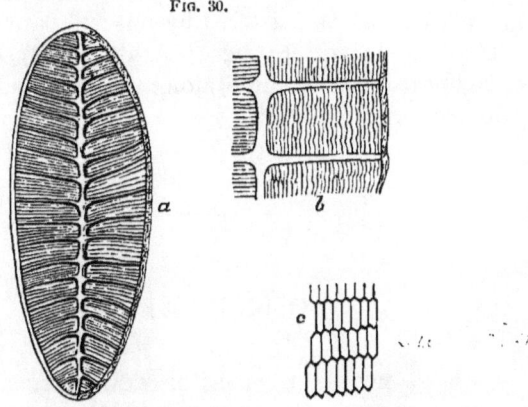

Fig. 30.

Valve of Surirella Gemma.
a. Transverse ridges. *b.* Longitudinal lines. *c.* The same, resolved into areolations.

Surirella gemma are estimated at 30 to 32 in $\tfrac{1}{100}$th of a millimetre, and the markings on *Grammataphora subtilissima* at 32 to 34 in the same distance.

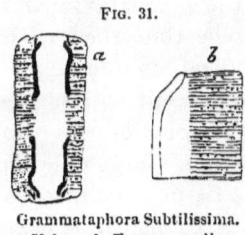

Fig. 31.

Grammataphora Subtilissima.
a. Valve. *b.* Transverse lines.

J. D. Moller has produced a very excellent test-plate, containing twenty diatoms, with descriptions, according to their value as tests.

The *Pleurosigma angulatum* (Plate I, Fig. 29), with suitable power and illumination, should show distinct hexagonal areolations. The *Surirella gemma* (Fig. 30) shows a series of fine transverse lines across the ridges which run from the edge to the central line. The finest of these ridges are not always readily seen, and the transverse ones are only to be mastered by toil and patience.

The *Grammataphora subtilissima* (Fig. 31) shows transverse lines (or rows of dots) along the edge, and sometimes a double series of oblique lines.

CHAPTER V.

MODERN METHODS OF EXAMINATION.

Microscopy does not limit its researches to optical enlargement, but seeks to comprehend elementary structure, and its methods vary according to the object immediately in view. It may seek merely to discern the form or morphology of the elementary parts or their peculiar functions. It may be concerned with inorganic forms, normal or pathological anatomy, or with physiology. Each department of pursuit will suggest some variation, yet a general plan of operation is possible.

Coarse, and moderately large objects, as a small insect, a piece of vegetable tissue, etc., may be observed by placing it in the forceps, or on the stage of the instrument, under an objective of low power, but ordinarily a considerable degree of preparation is needed in order to acquire a true idea of structure.

Most of the tissues to be examined are in a moist condition, and many require to be dissected or preserved in fluid. This has much to do with the appearance of the object in the microscope. If fibres or cells are imbedded in connective tissue or in fluids, of which the refractive power is the same as their own, they cannot be perceived even with the best glasses, and artificial means must be resorted to that they may become visible. The refractive power of different media causes different appearances. Thus a glass rod lying in water is easily seen, but in Canada balsam, whose refractive power is nearly the same as glass, it is barely seen as a flat band, and in the more highly refractive anise oil it presents the appearance of a cavity in the oil.

During life the cavities and fissures in animal tissues, in consequence of the different refractive power of their contents and the change which takes place soon after death, exhibit a sharpness and softness of outline which is seldom seen in preparations.

There are two methods of microscopic investigation or of preparation preliminary to direct observation: 1. Mechanical, for the separation and isolation of the elementary parts. 2. Chemical, which dissolve the connecting material, or act on it differently than on other elements.

For minute dissection a great variety of instruments have been proposed, but by practiced hands more can be accomplished in shorter time by simple means than with complicated ones. Two or three scalpels, or small anatomical knives, a pair of small scissors, such as is used in operations on the eye, and fine-pointed forceps, will be found useful. But the most serviceable instruments are dissecting-needles, such as the microscopist may make for himself. A common sewing-needle, with the eye end thrust into a cedar stick about three inches long and one-fourth of an inch diameter, will answer the purpose. The point should not project so far as to spring, and if desired, a cutting edge can be given to it by a hone.

The light should be concentrated on the work by means of a bull's-eye condenser, and as far as possible, the dissection should be carried on with the unassisted eye. Very often the work is so fine that a magnifying glass, or simple microscope, fixed to a suitable arm, will be needed. A large Coddington lens, an inch and a half in diameter, such as is used frequently by miners, will be useful. Sometimes it is necessary to resort to the dissecting microscope, which is a simple lens, of greater or less power, arranged with rack and pinion, mirror, etc.

The specimen may be dissected under water, in a glass or porcelain dish, or a trough made of gutta-percha, etc. Dr. Lawson's binocular dissecting microscope (Fig. 32) is

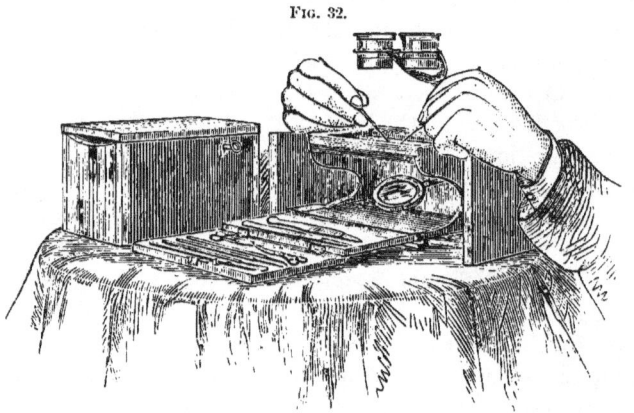

Fig. 32.

Lawson's Binocular Dissecting Microscope.

a most useful form, as both eyes may be used. Loaded corks, with sheet lead fastened to their under surface, may be used to pin the subject on for greater facility in dissection. Rests, or inclined planes of wood, one on each side of the trough, will give steadiness to the hands. Camels'-hair pencils for the removal of dust and extraneous elements, and for spreading out thin and delicate tissues or sections, are indispensable. Pipettes, or glass tubes, one end of which can be covered with the end of the finger, may serve to convey a drop of fluid or a small specimen from a bottle.

Preparation of Loose Textures.—If the formed elements of tissue do not combine in a solid mass, it is only necessary to place a small quantity on a glass slide and cover it with a plate of thin glass. If the elements are too close for clear definition under the microscope, a drop of fluid may be added. The nature of this fluid, however, is not a matter of indifference. Some elements are greatly changed by water, etc., and it becomes important to consider the fluid which is most indifferent. Glycerin and water, one part to nine of water, will serve well for most objects. Animal tissues are often best treated with aqueous humor, serum, or iodized serum. A weak solution of salt, 7.5 grains chloride of sodium to 1000 grains of distilled water, serves for many delicate structures. (See section on *Fluid Media*.)

Preparation by Teasing.—A minute fragment of tissue should be placed in a drop of fluid on a slide, and torn or unravelled by two sharp needles. This is accomplished more easily after maceration, and sometimes it is necessary to macerate in a substance which will dissolve the connecting material. This picking or teasing should be slowly and accurately performed. Beginners often fail of a good preparation by ceasing too soon, as well as by having too large a specimen. The most delicate manipulation is required to isolate nerve-cells and processes.

Preparation by Section.—A section of soft substance may be made with a sharp knife or scalpel, or with a pair of scissors curved on the upper side. A section cut with the latter will taper away at the edges so as to afford a view of its structure.

Valentin's double knife (Fig. 33) is used for soft tissues where only a moderate degree of thinness is needed. The blades should be wet, or the section made under water.

Soft substances often require hardening before sections can be made. The most simple and best method is that of freezing, by surrounding the specimen with a freezing mixture, when it may be cut with a cold knife. Small pieces of tissue may be hardened in absolute alcohol, frequently renewed. Chromic acid, in solution of one-fourth to two per cent., is often used for animal tissues, or bichromate of potash of the same strength. A solution of one-fifth to one-tenth per cent. of perosmic acid or of chloride of palladium is also recommended.

Soft tissues often require imbedding in a concentrated solution of gum or of wax, spermaceti, or paraffin tempered with oil. In this case sections may be made readily by means of a section-cutter. For imbedding in wax, etc.,

FIG. 33.

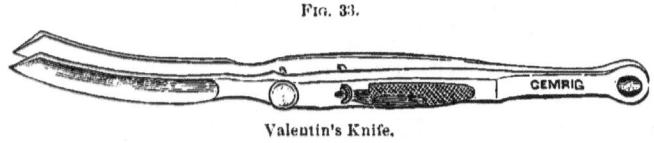

Valentin's Knife.

the specimen must be hardened in alcohol, then treated with oil of cloves or turpentine, and the section should be mounted in Canada balsam or Dammar varnish.

Sections of hard substances, or of those imbedded, are often made by machines invented for the purpose. One of the simplest is (Fig. 34) an upright hollow cylinder, with a kind of piston, pushed upwards by a fine screw. The upper end of the cylinder carrying the specimen terminates in a flat table, along which a sharp knife or flat razor is made to slide. At one side of the tube is a binding-screw for holding the specimen steady. A section may be cut by such an instrument after inserting the structure desired in a piece of carrot, etc., which may be placed in the tube; or the tube may be filled with wax, etc., and the specimen imbedded. Bones, teeth, shells, corals, minerals, etc., require to be cut with fine saws, or a disk of thin iron on a lapidary's wheel, and filed or ground down to the requisite thinness, then polished with emery, rouge, etc. The green oxide of chromium has been suggested to me as a useful polishing powder for hard substances. For calcareous substances, files and hones will suffice to reduce the thickness, and putty

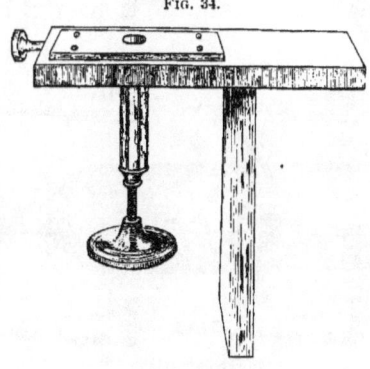

FIG. 34.

Section-Cutter.

powder or jewellers' rouge for polishing. They should be mounted in Canada balsam.

Staining Tissues.—Certain elements, not previously visible, can often be made evident by certain coloring matters, by which some constituents become more quickly or more thoroughly stained than others. The "germinal matter," or "bioplasm" of Dr. Beale, identical with the "protoplasm" or "sarcode" of other observers, may thus be distinguished from the "formed materials" or "tissue element," which are the products of its activity. Carmine, anilin, hæmatoxylin, and picric acid, are used for staining by imparting their own color to tissues; while nitrate of silver, chloride of gold, chloride of palladium, and perosmic acid stain, by their chemical action, often under the reducing influence of light. (See *Fluid Media*.)

Injecting Tissues.—Injections of the vessels in animal tissues are resorted to either to exhibit their course or the structure of the vascular walls. For the latter purpose a solution of nitrate of silver is commonly employed, for the former either opaque or transparent coloring matter. (See *Fluid Media*.)

The injecting syringe (Fig. 35) is made of brass or Ger-

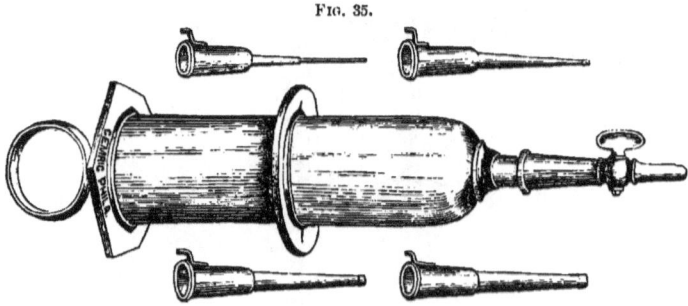

Fig. 35.

Injecting Syringe.

man silver. One of the pipes should be inserted into the principal vessel, as the aorta of a small animal, the umbilical vein of a fœtus, or the artery, etc., of an organ, and should be securely fastened by a thread. All other open vessels should be tied. The solution of gelatin, or other matter used, should be strained, so as to be free from foreign particles, and should be forced into the vessels with a gentle, steady pressure on the syringe.

Injections should be made soon after the death of the animal, or else after the rigor mortis has subsided.

Sometimes the syringe is substituted by a self-acting apparatus, consisting of a Wolfe's bottle, containing the fluid, which is pressed upon by a column of air from another source, and driven through a flexible tube to the pipe in the bloodvessel.

The older anatomists used colored plaster or wax to demonstrate the arteries and veins, but modern histology requires finer materials. Isinglass or gelatin, colored, and injected warm, or a solution of colored glycerin, are now resorted to. The former serves for opaque, and the latter for fine, transparent injections.

The art of injecting can only be learned by practice, yet perseverance, in despite of many failures, will insure success.

The liver, kidney, etc., may be injected separately, and it is often desirable to use various colors for the different sets of vessels. After injection thin slices may be cut off and mounted in fluid or balsam.

Preparation in Viscid Media.—Dr. Beale has proposed a method of preparing animal and vegetable tissues for examination with the very highest powers, which has led to valuable results. It consists in using pure glycerin or strong syrup, instead of watery solutions. In this way an amount of pressure may be applied to sections, in order to render them thin enough for examination, which would be destructive to specimens in water, while the preservative action of the media prevents change in the structure. It is necessary to soak the specimen some time, and the strength of the fluid should be gradually increased until the tissue is permeated by the strongest that can be obtained. Dr. Beale has found that minute dissection is much more readily performed in such fluids, and that even very hard textures, as bone and teeth, may be softened by them, especially if acetic acid is added, so as to permit thin sections to be made with a knife. He recommends vessels to be first injected, as with fine, transparent blue, and the germinal matter to be stained with carmine. A few drops of a solution of chromic acid, or bichromate of potash, so as to impart to the glycerin a pale straw color, serves to harden even the finest nerve-structures. Acetic acid, and other reagents also, are much more satisfactorily used with glycerin than with water. If syrup is used, camphor, carbolic acid, etc , must be employed to prevent the growth of fungi, but pure glycerin is free from this inconvenience.

A great advantage of this mode of investigation consists in the fact that a specimen thus prepared is already mounted, and needs but a proper cement to the glass cover and a finish to the slide, when it is ready for the cabinet.

FLUID MEDIA.

1. INDIFFERENT FLUIDS.

The vitreous humor, amniotic liquor, serum, etc., which form the usual fluids termed indifferent, always contain what Prof. Graham designated colloid and crystalloid substances. In 1000 parts there are about 4 parts of colloid (albumen) and 7.5 of crystalloid substance (chloride of sodium).

The iodine serum of Schultze consists of the amniotic fluid of the embryo of a ruminant, to which about 6 drops of tincture of iodine to the ounce is added A small piece of camphor will preserve this from decomposition a long time. A substitute for this is composed of 1 ounce of white of egg, 9 ounces of water, 2 scruples chloride of sodium, with the corresponding quantity of tincture of iodine.

2. CHEMICAL REAGENTS.

The greatest care should be used with these, that the instrument and glasses may be preserved. A small drop, applied by a glass rod drawn out to a point to the edge of the glass cover, will suffice in most cases.

Sulphuric Acid.—Concentrated is used to isolate the cells of horny structures, as hair, nails, etc. Dilute (1 part to 2–3 of water) gives to cellulose, previously dyed with iodine, a blue or purple color, and, when mixed with sugar, a rose-red to nitrogenous substances and bile. 0.1 to 1000 of water, at a temperature of 35–40° C., resolves connective tissue into gelatin and dissolves it, so as to be useful in isolating muscular fibres.

Nitric Acid.—Diluted with 4 or 5 parts water, separates the elementary parts of many vegetable and animal tissues when they are boiled or macerated in it. With chlorate of potash it is still more energetic, but caution is needed in its use.

Muriatic Acid, Strong.—Used for dissolving intercellular substance, as in the tubes of the kidney, etc. Dilute for dissolving calcareous matter.

Chromic acid, $\frac{1}{2}$ to 2 per cent. solution for hardening nerves, brain, etc.

Oxalic acid, to 15 parts water, causes connective tissue to swell and become transparent, while albuminoid elements are hardened. Preserves well delicate substances, as rods of retina, etc.

Acetic acid makes nuclei visible and connective tissue transparent, so as to exhibit muscles, nerves, etc., otherwise invisible.

Iodine (1 grain of iodine, 3 grains iodide of potassium, 1 ounce of water) turns starch blue and cellulose brown.

Caustic potash or soda renders many structures transparent.

Lime-water or *baryta-water* is used for investigating connective structures, especially tendon, as maceration enables the needle to divide its fibrilla.

Chloride of Sodium.—Solutions of this salt for indifferent media should always have some colloid, as albumen or gum-arabic added (7.5 grains in 1000 grains of water for delicate structures).

Bichromate of potash is used in stronger solution for the same purposes as chromic acid.

Müller's eye-fluid for hardening the retina, and preserving delicate embryos, etc., consists of bichromate of potass., 2 grammes; sulphate of soda, 1 gramme; distilled water, 100 grammes.

Alcohol dissolves resins and many vegetable coloring matters; renders most vegetable preparations more transparent, and albuminous animal tissues more opaque.

Acetic acid and alcohol, 1 part of each to 2 of water, renders connective tissue transparent, and albuminoid tissue prominent. The proportions can be varied.

Alcohol and soda (8–10 drops of strong solution of caustic soda to each ounce) renders many tissues very hard and transparent. Beale recommends it for embryonic structures.

Ether dissolves resins, oils, and fat.

Turpentine renders dried animal sections transparent.

Oil of cloves acts as turpentine.

Solution of chloride of zinc, iodine, and iodide of potassium, is recommended by Schacht as a substitute for iodine and sulphuric acid to color vegetable cells, etc. Zinc is dissolved in hydrochloric acid, and the solution is evaporated to syrupy consistence in contact with metallic zinc. This is saturated with iodide of potassium, iodine added, and the solution diluted with water. Wood cells, after boiling in caustic potash, are stained blue by it.

Boracic acid, used by Prof. Brucke to separate the elements of red blood-corpuscles.

3. Staining Fluids.

Thiersch's Carmine Fluids.

a. Red Fluid.

1. Carmine, ... 1 part.
   Caustic ammonia, ... 1 "
   Distilled water, ... 3 parts. Filter.
2. Oxalic acid, ... 1 part.
   Distilled water, ... 22 parts.

1 part of carmine solution to 8 parts of the acid solution, add 12 parts absolute alcohol. Filter. After staining wash in 80 per cent. alcohol.

b. Lilac Fluid.

Borax, ... 4 parts.
Distilled water, ... 56 "
Dissolve and add,
Carmine, ... 1 part.

Mix with twice the volume of absolute alcohol and filter.

Beale's Carmine Fluids.

Carmine, ... 10 grains.
Strong liquor ammonia, ... ½ drachm.
Glycerin, ... 2 ounces.
Distilled water, ... 2 "
Alcohol, ... ½ ounce.

Dissolve the carmine in the ammonia in a test-tube by aid of heat; boil it and cool and add the other ingredients. Filter.

Acid Carmine Fluid.—Mix ammoniacal solution of carmine with acetic acid in excess and filter. This is said to stain diffusely, but adding glycerin with muriatic acid (1 : 200), concentrates the color in the cell-nucleus.

Anilin (or Magenta) Red Fluid.

Fuchsin (crystal), ... 1 centigramme.
Absolute alcohol, ... 20–25 drops.
Distilled water, ... 15 cubic centim.

Anilin Blue Fluid.—Anilin blue, treated with sulphuric acid and dissolved in water till a deep cobalt color is obtained.

Blue Fluid from Indigo Carmine.

Oxalic acid,	1 part.
Distilled water,	20–30 parts.
Indigo carmine to saturation.	

Logwood Violet Fluid.

1. Hæmatoxylin, ... 20 grains.
   Absolute alcohol, ... ½ ounce.
2. Solution of 2 grains of alum to 1 ounce of water.

A few drops of the first solution to a little of the second in a watch-glass, etc.

Picro-Carmine Fluid.—Filter a saturated solution of picric acid, and add, drop by drop, strong ammoniacal solution of carmine till neutralized.

Nitrate of Silver Fluid.—Fresh membranous tissues, exposed to 0.5 to 0.2 per cent. solution of nitrate of silver, washed and exposed to light, often show a mosaic of epithelium, etc.

Osmic Acid.—$\frac{1}{10}$th to 1 per cent. solution stains the medulla of nerves, etc., black.

Chloride of Gold.—The solution should be similar to that of nitrate of silver. Exposure to light stains the nerves, etc., a violet or red color.

Prussian Blue.—After immersing a tissue in 0.5 to 1 per cent. solution of a protosalt of iron, dip it in a 1 per cent. solution of ferrocyanide of potassium.

Other Staining Fluids.—Marked effects are often produced by the use of the violet, blue, and other inks in the market. Thus I succeeded in some demonstrations of nerve plexuses in muscle better than in any other way. I suspect the particular ink employed contained a large per cent. of soluble Prussian blue.

4. Injecting Fluids.

For opaque injection several plans have been devised. Resinous and gelatinous substances, variously colored, are most usual. Lieberkuhn used tallow, varnish, and turpentine, colored with cinnabar; and Hyrtl, whose preparations have been much admired, follows a similar plan. He evaporates pure copal or mastic varnish to the consistence of syrup, and grinds one-eighth as much cinnabar and a little wax with it on a slab. For fine injections this is diluted with ether.

For a bright red, the cinnabar may be mixed with a little carmine

For a yellow color, the chromate of lead, prepared by mixing solutions of acetate of lead (30 parts to 2 ounces of water), and red chromate of potash (15 parts).

White may be made with zinc-white or carbonate of lead—4¼ ounces of acetate of lead in 16 ounces of water, mixed with 3¼ ounces carbonate of soda in 16 ounces.

For gelatinous injections the coloring matter is combined with jelly, prepared by soaking fine gelatin in cold water for several hours, then dissolving in a water-bath and filtering through flannel.

By injecting gelatinous fluid solutions of various salts, the coloring matter may be left in the vessels by double decomposition.

A red precipitate, with iodide of potassium and bichloride of mercury.

A blue, by ferrocyanide of potassium and peroxide of iron, etc.

Dr. Goadby's formula for a yellow color is:

Saturated solution of bichromate of potassium, .	8 ounces.
Water, ...	8 "
Gelatin, ...	2 "
Saturated solution of acetate of lead, . . .	8 ounces.
Water, ...	8 "
Gelatin, ...	2 "

For gelatinous injections, both the fluid and the subject should be as warm as may consist with convenience. Camphor also should be added to prevent mould.

For transparent injections, gelatin may be used combined with colored solutions, or still better, glycerin, which may be used cold.

Thiersch's Blue.—Half an ounce of warm concentrated solution (2:1) of fine gelatin is mixed with 6 cubic centimetres of a saturated solution of sulphate of iron. In another vessel, 1 ounce of the gelatin solution is mixed with 12 cubic centimetres of saturated solution of ferrocyanide of potassium, to which 12 cubic centimetres of saturated solution of oxalic acid is added. When cold, add the gelatinous solution of sulphate of iron drop by drop, with constant stirring, to the other. Warm it, still stirring, and filter through flannel.

Gerlach's Carmine.—Dissolve 5 grammes (77 grains) of fine carmine in 4 grammes (70 grains) of water and $\frac{1}{2}$ gramme (8 drops) of liquor ammonia. Let it stand several days (not airtight), and mix with a solution of 6 grammes of fine gelatin to 8 grammes of water, to which a few drops of acetic acid are added.

Thiersch's Yellow.—Prepare a solution of chromate of potash (1:11), and a second solution of nitrate of lead, of same strength. To 1 part of the first add 4 parts of solution of gelatin (about 20 cubic centimetres to 80), and to 2 parts of the second add 4 parts of gelatin (40 cubic centimetres to 80). Mix slowly and carefully, heat on a water-bath, and filter through flannel.

Equal parts of Thiersch's blue and yellow carefully mixed and filtered make a good green.

COLD TRANSPARENT INJECTIONS.

Beale's Blue.

Glycerin,	1 ounce.
Alcohol,	1 "
Ferrocyanide of potassium,	12 grains.
Tincture of perchloride of iron,	1 drachm.
Water,	4 ounces.

Dissolve the ferrocyanide in 1 ounce of water and glycerin, and the muriated tincture of iron in another ounce. Add the latter very gradually to the other, shaking often; then gradually add the alcohol and water.

Beale's Finest Blue.

Price's glycerin,	2 ounces.
Tincture of perchloride of iron,	10 drops.
Ferrocyanide of potassium,	3 grains.
Strong hydrochloric acid,	3 drops.
Water,	1 ounce.

Mix the tincture of iron with 1 ounce glycerin and the ferrocyanide, after dissolving in a little water, with the other ounce. Add the iron to the other solution gradually, shaking well. Lastly, add the water and hydrochloric acid. Sometimes about 2 drachms of alcohol are added.

Müller's Blue.—This is made by precipitation of soluble Prussian blue from a concentrated solution by means of 90 per cent. alcohol.

Beale's Carmine.—Mix 5 grains of carmine with a few drops of water, and when well incorporated, add 5 or 6 drops of liquor ammonia. To this add ½ ounce of glycerin, and shake well. Another ½ ounce of glycerin containing 8 or 10 drops of acetic or hydrochloric acid is gradually added. It is then diluted with ½ ounce of glycerin, 2 drachms of alcohol, and 6 drachms of water.

Nitrate of Silver Injection.—For demonstrating the structure of the bloodvessels, the animal is bled, and a solution of 0.25 to 1 per cent. of nitrate of silver, or a mixture of gelatin with such a solution, is used.

5. Preservative Fluids.

Canada Balsam.—This is perhaps the most common medium used. When an object is not very transparent, and drying will not injure it, balsam will do very well, but it is not adapted to moist preparations. Colonel Woodward, of Washington, uses a solution of dried or evaporated Canada balsam in chloroform or benzole.

Dammar Varnish.—Dr. Klein and other German histologists prefers this to Canada balsam. Dissolve ½ to 1 ounce of gum Dammar in 1 ounce of turpentine; also ½ to 1 ounce of mastic in 2 ounces of chloroform. Mix and filter.

Glycerin.—This fluid is universally useful to the microscopist. (See *Preparation in Viscid Media*, page 65.) Vegetable and animal substances may be preserved in glycerin, but if it is diluted, camphor or creasote must be added to prevent confervoid growths. It is said, however, to dissolve carbonate of lime.

Gelatin and Glycerin.—Soak gelatin in cold water till soft, then melt in warm water, and add an equal quantity of glycerin.

Gum and Glycerin.—Dissolve 1½ grains of arsenious acid in 1 ounce of water, then 1 ounce of pure gum arabic (without heat), and add 1 ounce of glycerin.

Deane's Compound.—Soak 1 ounce of gelatin in 5 ounces of water till soft; add 5 ounces of honey at a boiling heat. Boil the mixture, and when cool, add 6 drops of creasote in ½ ounce of alcohol; filter through flannel. To be used warm.

Carbolic Acid.—1 : 100 of water is a good preservative.

Thwaite's Fluid.—To 16 parts of distilled water, add 1 part of rectified spirit and a few drops of creasote; stir in a little prepared chalk, and filter. Mix an equal measure of camphor-water, and strain before using. For preservation of algæ.

Solution of Naphtha and Creasote.—Mix 3 drachms of creasote with 6 ounces of wood naphtha; make a thick, smooth paste with prepared chalk, and add gradually, rubbing in a mortar, 64 ounces of water. Add a few lumps of camphor, and let it stand several weeks before pouring off or filtering the clear fluid. Dr. Beale recommends this highly for the preservation of dissections of nerves and morbid specimens.

Ralf's Fluid.—As a substitute for Thwaite's fluid in the preservation of algæ. 1 grain of alum and 1 of bay salt to 1 ounce of distilled water.

Goadby's Solution.—Bay salt, 4 ounces; alum, 2 ounces; corrosive sublimate, 4 grains; boiling water, 4 pints. This is the strength most generally useful, although it may be made stronger or more dilute. It is a useful fluid. If the specimen contain carbonate of lime, the alum must be left out, and the quantity of salt may be quadrupled.

Dr. Beale discards all solutions containing salts for microscopic purposes, as they render the textures opaque and granular.

Soluble Glass, or a solution of silicate of soda or potash, or of both, has been proposed, but it is apt to render specimens opaque.

Chloride of Calcium in saturated aqueous solution has been much recommended, especially by botanists.

Acetate of Potash, a nearly saturated solution, is useful for vegetable preparations and for specimens of animal tissue which have been stained with osmic acid. The latter do not bear glycerin.

Pacinian Fluid.—This is variously modified, but may consist of corrosive sublimate, 1 part; chloride of sodium, 2 parts; glycerin, 13 parts; distilled water, 113 parts. Sometimes acetic acid is substituted for chloride of sodium.

6. Cements.

Gold Size is recommended by Dr. Carpenter as most generally useful for thin covers. It is made by boiling 25 parts of linseed oil for three hours with 1 part of red lead and ½ of as much umber. The fluid part is then mixed with yellow ochre and white lead in equal parts, so as to thicken it, the whole boiled again, and the fluid poured off for use.

Bell's Cement is said to be best for glycerin specimens. It appears to be shellac dissolved in strong alcohol.

Brunswick Black is asphaltum dissolved in turpentine. A little india-rubber dissolved in mineral naphtha is sometimes added.

Canada Balsam in chloroform or Dammar varnish (page 74) is often used as a cement.

Marine Glue.—This is most useful in building glass cells, etc. It consists of equal parts of shellac and india-rubber dissolved in mineral naphtha by means of heat.

Electrical Cement is made by melting together 5 parts of rosin, 1 of beeswax, and 1 of red ochre. 2 parts of Canada balsam added make it more adhesive to glass.

White, hard Varnish, or gum sandarac, dissolved in alcohol and mixed with turpentine varnish, is sometimes colored by lampblack, sealing-wax, etc.

White Zinc Cement.—Oxide of zinc rubbed up with equal parts of oil of turpentine and 8 parts of solution of gum Dammar in turpentine of a syrupy consistence, or Canada balsam, chloroform, and oxide of zinc.

CHAPTER VI.

MOUNTING AND PRESERVING OBJECTS FOR THE MICROSCOPE.

For the permanent preservation of specimens, various means are employed, according to the nature of the object and the particular line of investigation desired. Few, if any, objects show all their peculiarities of structure or adaptation to function, and for scientific work it is often

MOUNTING AND PRESERVING OBJECTS. 77

necessary to have the same structure prepared in different ways.

Opaque Objects have sometimes been attached by thick gum to small disks of paper, etc., or to the bottom and sides of small pill-boxes, or in cavities in slides of bone or wood, but they are better preserved on glass slides, as hereafter described.

The most convenient form of slide for microscopic purposes is made of flattened crown or flint glass, cut into slips of three inches by one inch, and ground at the edges. Some preparations are mounted on smaller slips, but they are less convenient than the above, which is regarded as the standard size.

On such slides objects are fixed, and covered by a square or round piece of thin glass, varying from $\frac{1}{20}$th to $\frac{1}{250}$th of an inch in thickness. Both slides and thin glass can be procured at opticians' stores. Laminæ of mica or talc are sometimes used for lack of better material, but are too soft. For object-glasses of the shortest focal length, however, it is necessary at times to resort to this sort of covering.

Great care should be taken to have both slide and cover clean. With thin glass this is difficult, owing to its brittleness. Practice will teach much, but for the thinnest glass two flat pieces of wood covered with chamois leather, between which the cover may lie flat as it is rubbed, will be serviceable.

Very thin specimens may be mounted in balsam, glycerin, etc., covered with the thin glass cover, and then secured by a careful application of cement to the edges of the cover. If, however, the pressure of the thin glass be objectionable, or the object be of moderate thickness, some sort of cell should be constructed on the slide.

The thinnest cells are made with cement, as gold size, Brunswick black, etc., painted on with a camel's-hair pencil. For preparing these with elegance, Shadbolt's *turn-*

table has been contrived (Fig. 36). The slide is placed between the springs, and while rotated, a ring of varnish of suitable breadth is made on the glass.

A piece of thin glass (or even of thick glass) may be perforated and cemented to the slide with marine glue by

Shadbolt's Turntable for making Cement-Cells.

the aid of heat; or vulcanite, lead, tin, gutta percha, etc., may be made into a cell in a similar way as seen in Fig. 37.

The perforation of thin glass may be easily performed by cementing it over a hole in a brass plate, etc., with marine glue, and punching it through with the end of a

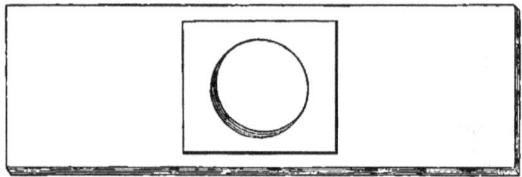

Cell of Glass, Vulcanite, etc.

file. The edges may then be filed to the size of the hole, and the glass removed by heating the brass. Thicker glass may be bored with a file by moistening it with turpentine.

Dry objects, especially if they are transparent, as diatoms, thin sections of bone, crystals, etc., may be attached to the slide with Canada balsam, etc., covered with thin glass, which should be cemented at the edges, and gummed over all a strip of colored or lithographed paper, in which an aperture has been made with a punch.

Mounting in Balsam or Dammar Varnish is suitable for a very large proportion of objects. It increases the transparency of many structures, filling up interstices and cavities, and giving them a smooth, beautiful appearance. Very delicate tissues, as fine nerves, etc., are rendered invisible by it, and require other fluids, as glycerin.

Before mounting in balsam, the object should be thoroughly dry, otherwise a milky appearance will result. It should then be placed in oil of cloves or of turpentine to remove greasiness and increase the transparency. A clean slide, warmed over a spirit-lamp or on a hot plate, should then have a little balsam placed on its centre, and the object carefully lifted from the turpentine is put into the balsam and covered with another drop. The slide should then be gently warmed, and any air-bubbles pricked with a needle-point or drawn aside. The thin glass cover should be warmed and put on gently, in such a way as to lean first on one edge and fall gradually to a horizontal position. The slide may be warmed again, and the superfluous balsam pressed from under the cover by the pressure of a clean point upon it.

If the object is full of large air-spaces and is not likely to be injured by heat, the air may be expelled by gently boiling it in the balsam on the slide. If the object be one which will curl up, or is otherwise injured by heat, the air-pump must be resorted to. A cheap substitute for the air-pump may be made by fitting a piston into a tolerably wide glass tube closed at one end. The piston should have a valve opening outwards. The preparation in balsam may be placed at the bottom of the tube, and a few strokes of the piston will exhaust the air.

To fill a deep cell with Canada balsam, it may be well to fill it first with turpentine and place the specimen in it. Then pour in the balsam at one end, the slide being inclined so that the turpentine may run out at the other. Lay the cover on one edge of the cell and gradually lower it till it lies flat. In this way air may be excluded.

The solution of balsam in chloroform needs no heat, and has little liability of air-bubbles.

The excess of balsam round the edge of the glass cover may be removed with a knife and cleaned with turpentine or benzine, etc.

For animal tissues, the oil of cloves is sometimes used instead of turpentine to increase the transparency, and a wet preparation, as a stained or injected specimen, may be mounted in balsam or Dammar by first placing it in absolute alcohol to extract the water, then transferring to oil of cloves or turpentine, and lastly, to the balsam. In a reverse order, a specimen from balsam may be cleaned and mounted in fluid.

Mounting in Fluid is necessary for the preservation of the most delicate tissues and such as may be injured by

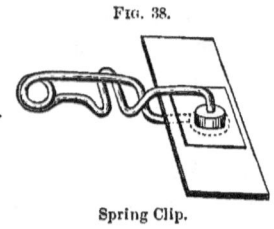

Fig. 38.

Spring Clip.

drying. Glycerin is perhaps the most generally useful fluid. (See *Preservative Fluids*, page 73.)

For mounting in fluid, it is safer to have a thin cell of varnish prepared first than to risk the running in of the cement under the cover, as will be likely to occur otherwise.

The air-pump is sometimes needed in mounting in fluid to get rid of air-bubbles. A spring clip (Fig. 38) is also a useful instrument for making moderate pressure on the glass cover until the cement on its edge is dry. A dropping-tube with a bulbous funnel, covered with thin india-rubber, for taking up and dropping small quantities of fluid, will also be of service.

Superfluous fluid may be removed from the edge of the cover by a piece of blotting-paper, care being used not to draw away the fluid beneath the cover.

As soon as objects are mounted, the slides should be labelled before cementing is finished, otherwise time will be lost in searching for a particular object among others, or the name may be forgotten.

Boxes of wood or of pasteboard, with grooved racks at the sides, are occasionally used for preserving a collection of specimens. It is better, however, to have a cabinet with drawers or trays so that the specimens may lie flat, with their ends towards the front of the drawer. A piece of porcelain on the end of the drawer is convenient for the name of the class of objects contained, to be written on with lead-pencil.

Collecting Objects.—The methods pursued by naturalists generally will suffice for a large proportion of the objects which are matters of microscopic inquiry, but there are others which, from their minuteness, require special search. Many fresh-water species of microscopic organisms inhabit pools, ditches, and streams. Some attach themselves to the stems and leaves of aquatic plants, or to floating and decaying sticks, etc. Others live in the muddy sediment at the bottom of the water. A *pond stick* has been contrived for the collection of such organisms, consisting of two lengths, sliding one within the other, so that it may be used as a walking-cane. In a screw socket at one end may be placed a curved knife for cutting portions of plants which contain microscopic parasites; or a screw collar for carrying a screw-topped bottle, which serves to bring up a sample of liquid; or it may have a ring for a muslin net.

The net should be confined by an india-rubber band in a groove, so as to be slipped off readily and emptied into a bottle. The collector should have enough bottles to keep organisms from each locality separate, and when animalcules are secured enough, air should be left to insure their safety.

Marine organisms may be obtained in a similar way if they inhabit the neighborhood of the shore, but others can only be secured by means of the dredge or tow-net. The latter may be of fine muslin sewn to a wire ring of twelve inches diameter. It may be fastened with strings to the stern of a boat, or held by a stick so as to project from the side. For the more delicate organisms, the boat should be rowed slowly, so that the net may move gently through the water. Firmer structures may be obtained by attaching a wide-mouthed bottle to the end of a net made conical, and double, so that the inner cone may act as a valve. The bottle may be kept from sinking by a piece of cork. Such a net may be fixed to the stern of a vessel, and drawn up from time to time for examination.

Minute organisms may be examined on the spot by fishing them out of the bottle with a pipette, or small glass tube, and placing them on a slide. A Coddington or other pocket lens will suffice to show which are desirable for preservation.

Many of the lower animals and plants may be kept alive in glass jars for some time. Frogs, etc., may be kept under wire covers with a large piece of moist sponge.

Aquaria of various sorts may be procured and stocked at small expense, and will afford a constant source of instruction. For fresh-water aquaria the bottom of the jar, etc., should be covered with rich black earth, made into a paste, and this should be surmounted with a layer of fine washed sand. Roots of *Valisneria*, *Anacharis*, or *Chara* may then be planted in the earth and the vessel filled with water. As soon as the water is clear, put a few fresh-water molluscs in to keep down the growth of confervæ, especially such as feed on decayed vegetable matter, as *Planorbis carinatus*, *Paludina vivipara*, or *Amphibia glutinosa*. When bubbles of oxygen gas appear, fish, water insects, etc., may be introduced.

Marine aquaria require more skill than those for fresh water, but for temporary purposes, the plan described by Mr. Highley, in Dr. Beale's *How to Work with the Microscope*, is excellent. He fills a number of German beaker glasses with fresh sea-water, and places them in a sunny window. He then drops in each one or two limpet shells from which the animals have been removed, and upon which small plants of *Enteromorpha* and *Ulva* are growing. In a short time the sides of the jars next the light become coated with spores. He keeps the other sides clean with a piece of wood or sponge, so as to observe the small marine animals which may now be placed in the beakers. In this way a collection will keep healthy for months. After the sides are covered with spores, the sea-weeds may be removed, and the jars placed on a table at such a distance from the window that the light impinges only on the coated half, taking care that there is sufficient light to stimulate the spores to throw off bubbles of oxygen daily.

Prawns, fish, actiniæ, etc., may be fed on shreds of beef which has been pounded and dried, and then macerated in sea-water for a few minutes. All dead animals, slime, or effete matter should be removed by wooden forceps, etc., as soon as noticed.

CHAPTER VII.

THE MICROSCOPE IN MINERALOGY AND GEOLOGY.

Microscopic examination of minute fossil organisms, as Diatoms, Foraminifera, spicules of sponge, etc., has long been a subject of interest. Latterly, however, the microscope has been found to be essential to the study of physical geology and petrology. How many crude and verbose theories respecting cosmogony will disappear by this means of investigation time must reveal, but the animal nature of the *Eozoon Canadense* found in the Serpentine Limestone of the Laurentian formation of Canada, parallel with the Fundamental Gneiss of Europe, and the discovery by Mr. Sorby[*] of minute cavities filled with fluid in quartz and volcanic rocks, are indications that speculations based upon a merely external or even chemical examination of rock structures are immature and inadequate.

The systematic study of microscopic mineralogy and geology will require a large outlay of time and patience, and the field is one which is scarcely trodden. The plan of this work will only permit a brief outline, sufficient to aid a beginner, and indicating the value and the methods of minute investigation.

Preparation of Specimens.—Examination of the outer surface of a mineral specimen, viewed as an opaque body with a low power and by condensed light, is sometimes useful. The metals and their alloys, with most of their combinations with sulphur, etc , admit of no other method. Occasionally, as in iron and steel, the microscopic structure is best seen by polishing the surface, and then allowing the action of very dilute nitric acid. Mr. Forbes[†] states

---

[*] See Beale's How to Work with the Microscope.
[†] The Microscope in Geology, Popular Science Review, No. 25.

that many vitreous specimens (quite transparent) show no trace of structure until the surface has been carefully acted on by hydrofluoric acid.

It is generally necessary to have the specimens flat and smooth, and thin enough to transmit light. Sometimes fragments may be thin enough to show structure when mounted in balsam, as in the case of quartz, obsidian, pitchstone, etc., but usually thin sections must be ground and polished.

Chip off a fragment of the rock as flat and thin as possible, or cut with a lapidary's wheel, or a toothless saw of sheet-iron with emery. Grind down the specimen on an iron or pewter plate in a lathe until perfectly flat. Then grind with finer emery on a slab of fine-grained marble or slate, and finish with water on a fine hone, avoiding all polishing powders or oil. When perfectly smooth, cement the specimen on a square of glass with Canada balsam, and grind the other side until as thin as necessary, finish as before, remove it from the glass, and mount on a glass slide in balsam.

In this way, most silicates, chlorides, fluorides, carbonates, sulphates, borates, many oxides, sulphides, etc., may be prepared for examination by transmitted light. Very soft rocks may be soaked in turpentine, then in soft balsam, and afterwards heated until quite hard. The deep scratches on hard minerals, like quartz, left by the use of coarse emery, may be removed by using fine emery paper held flat on a piece of plate glass, and finally polished with rouge on parchment. Perhaps oxide of chromium from its hardness will be found the best polishing material. Crystals of soluble salts may be ground on emery paper and polished with rouge. Sometimes much may be learned by acting on one side only of a specimen with dilute acid.

Examination of Specimens.—The object of microscopic examination of minerals is to determine not only the nature of the material of which they are composed, but also, and chiefly, their structure, whether homogeneous, derived from the débris of previous rocks, or from the agency of the organic world. Ordinary mineralogical characteristics, as to hardness, specific gravity, color, lustre, form, cleavage, and fusibility, and above all, chemical composition, may suffice to show the material, but the microscope will give valuable assistance to this end, and is essential to a knowledge of structure.

Crystalline Forms.—The laws of crystallography teach that each chemical combination corresponds to a distinct relation of all the angles which can possibly arise from the primary form, so that the angular inclination of the facets of a crystal is a question of importance. This can be ascertained by a microscope having a revolving stage, properly graduated, or by the use of a *goniometer*, which is a thread stretched across the focus of the eye-lens, and attached to a movable graduated circle and vernier. The eye-piece attached to the polariscope of Hartnack is thus arranged, so as to act also as a goniometer.

Crystals are assumed to possess certain axes, and the form is determined by the relation of the plane surface to these axes. Although the forms of crystals are almost infinitely varied, they may be classified into seven crystallographic systems.

1. *The Regular Cubic or Monometric System* (Fig. 39)— These crystals are symmetrical, about three rectangular axes. The simplest forms are the cube and octahedron. Examples, diamond, most metals, chloride of sodium, fluor spar, alum.

2. *The Quadratic or Dimetric System* (Fig. 40).—Crystals symmetrical, about three rectangular axes, but only two axes of equal length. Examples, sulphate of nickel, tungstate of lead, and double chloride of potassium and copper.

3. *Hexagonal or Rhombohedral System* (Fig. 41).—Crystals with four axes; three equal in length, in one plane,

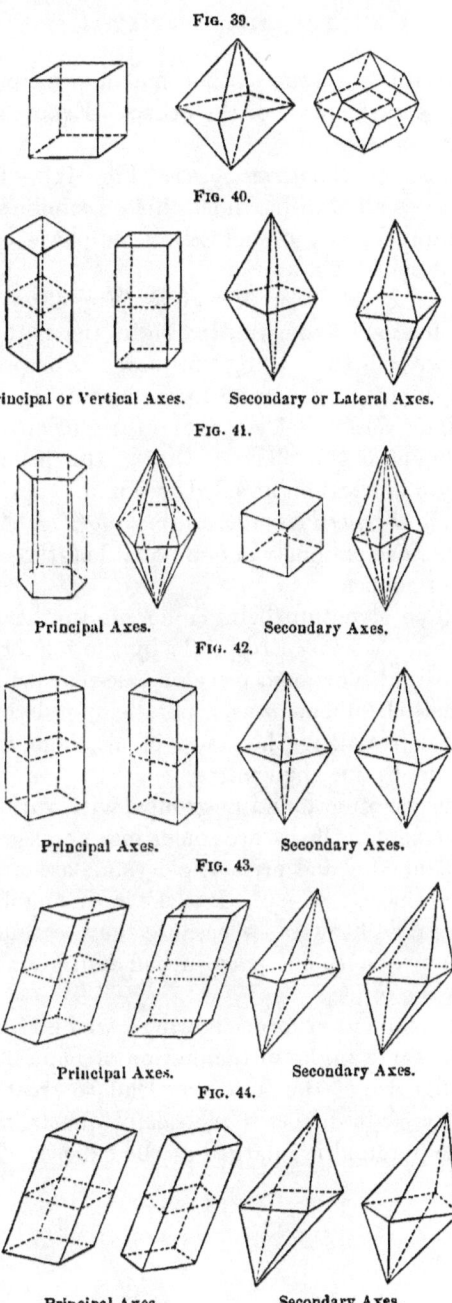

FIG. 39.

FIG. 40.

Principal or Vertical Axes.   Secondary or Lateral Axes.

FIG. 41.

Principal Axes.   Secondary Axes.

FIG. 42.

Principal Axes.   Secondary Axes.

FIG. 43.

Principal Axes.   Secondary Axes.

FIG. 44.

Principal Axes.   Secondary Axes.

and inclined 60° to each other, and a principal axis at right angles to the plane of the others. Examples, quartz, beryl, and calc-spar.

4. *Rhombic or Trimetric System* (Fig. 42).—Three rectangular axes, all of different lengths. Examples, sulphate of potassium, nitrate of potassium, sulphate of barium, and sulphate of magnesium.

5. *Oblique Prismatic or Monoclinic* (Fig. 43).—Two axes obliquely inclined, and a third at right angles to the plane of these two; all three being unequal. Examples, ferrous sulphate, sugar, gypsum, and tartaric acid.

6. *Diclinic System.*—Two axes at right angles, and a third oblique to the plane of these; the primary form being a symmetrical eight sided pyramid.

7. *Doubly Oblique Prismatic or Triclinic* (Fig. 44).—Three axes all inclined obliquely and of equal length. Example, sulphate of copper.

Crystalline structure being inherent in the nature of the mineral, becomes perceptible by the manner of division. A slight blow on a piece of calc-spar will separate it into small rhombohedrons or parallelopipeds, or produce internal fissures along the planes of cleavage, which will suffice to determine their angles.

Crystals are often found in groups, with various modes of arrangement. Cubes are sometimes aggregated so as to form octahedra, and prismatic crystals are often united together at one extremity. But the most singular groups are those called hemitropes, because they resemble a crystal cut in two, with one part turned half round and re-united to the other.

In all the numerous forms, however, we find in the same species the same angles or inclination of planes, although the unequal size of the faces may lead to great apparent irregularity, as in distorted crystals of quartz, where one face of the pyramid is enlarged at the expense of the rest.

An apparent distortion may also be produced by an oblique section.

The following examples may be of service, as showing the value of angular measurement in minerals:

Quartz. Rhombohedral system. Inclination of two adjoining faces 94° 15′.

Felspar. Monoclinic. Cleavage planes at right angles.

Albite or soda felspar. Triclinic. Angle 93° 36′.

Mica. Oblique prisms.

Magnesian mica. Right, rhombic, or hexagonal prisms.

Garnet. Dodecahedrons or trapezohedrons.

Idocrase. Square prisms.

Epidote. Oblique prisms.

Scapolite. Square and octagonal prisms.

Andalusite. Prisms of 90° 44′.

Staurotide. Rhombic prisms of 129° 20′.

Tourmaline. Three, six, nine, or twelve-sided prisms.

Topaz. Rhombic prisms of 124° 19′.

Beryl. Six-sided prisms.

Hornblende. Monoclinic. 124° 30′.

Augite or pyroxene. Monoclinic. 87° 5′.

Calcite or carbonate of lime. Forms various, but 105° 5′ between the cleavage faces.

Magnesite. Angle 107° 29′.

Dolomite. 106° 15′.

Gypsum. Monoclinic.

Crystals within Crystals.—Many specimens which appear perfectly homogeneous to the naked eye are shown by the microscope to be very complex. The minerals of erupted lavas are often full of minute crystals, leading to very anomalous results of chemical analysis. Some care is needed at times to distinguish such included minerals from cavities filled with fluid. The use of polarized light will sometimes determine this point.

Cavities in Crystals.—Mr. Sorby has shown that the various cavities in minerals containing air, water, glass, or stone will often indicate under what conditions the rock was formed. Thus crystals with water cavities were formed from solution; those with stone or glass cavities from igneous fusion; those with both kinds by the combined influence of highly heated water and melted rock under great pressure; while those that contain no cavities were formed very slowly, or from the fusion of homogeneous substance.

Use of Polarized Light.—Mr. Sorby states that the action of crystals on polarized light is due to their double refraction, which depolarizes the polarized beam, and gives rise to colors by interference if the crystal be not too thick in proportion to the intensity of its power of double refraction. This varies much, according to the position in which the crystal is cut, yet we may form a general opinion, since it is the intensity and not the character of the depolarized light which varies according to the position of the crystal. There are two axes at right angles to each other, and when either of them is parallel to the plane of polarization, the crystal has no depolarizing action, and if the polarizing and analyzing prisms are crossed, it looks black. On rotating the crystal or the plane of polarization, the intensity of depolarizing action increases until the axes are at 45°, and then diminishes till the other axis is in the plane. If, therefore, this takes place uniformly over a specimen, we know that it has one simple crystalline structure, but if it breaks up into detached parts, we know it is made up of a number of separate crystalline portions.

The definite order that may occur in the arrangement of a number of crystals may indicate important differences. Some round bodies, for example, like oolitic grains, have been formed by crystals radiating from a common nucleus; whilst others, as in meteorites, have the structure of round bodies which crystallized afterwards.

Sir D. Brewster discovered that many crystals have

two axes of double refraction, or rather axes around which double refraction occurs. Thus nitre crystallizes in six-sided prisms, with angles of about 120°. It has two axes of double refraction inclined about $2\frac{1}{2}°$ to the axes of the prism, and 5° to each other, so that a piece cut from such a crystal perpendicular to the axes, shows a double system of rings when a ray of polarized light is transmitted. When the line connecting the axes is inclined 45° to the plane of polarization, a cross is seen, which gradually assumes the form of two hyperbolic curves on rotating the specimen If the analyzer be revolved, the black cross will be replaced by white, the red rings by green, the yellow by indigo, etc. These rings have the same colors as thin plates, or a system of rings round one axis. Mica has two sets of rings, with the angle between the axes of 60° to 75°. Magnesian mica gives an angle of 5° to 20°.

Determination of the Origin of Rock Specimens.—Mr. Forbes has shown that the primary or eruptive rocks, consisting chiefly of crystallized silicates, with small quantities of other minerals, are developed as more or less perfect crystals at all angles to one another, indicating the fluid state of the mass at some previous time. The secondary or sedimentary rocks consist of rocks formed by the immediate products of the breaking up of eruptive rocks, or are built of the débris of previous eruptive or sedimentary rocks, or composed of extracts from aqueous solution by crystallization, precipitation, or the action of organic life. The accompanying figures, selected from Mr. Forbes's article in the *Popular Science Review*, well illustrate this method of investigation. Plate II, Fig. 45, is a section of lava from Vesuvius, magnified twelve diameters, showing crystals of augite in a hard gray rock. Plate II, Fig. 46, is a volcanic rock from Tahiti, consisting of felspar, with olivine and magnetic oxide of iron, and numerous crystals of a pyroxenic mineral. Plate II, Fig. 47, is pitchstone from a dyke in new red sandstone, magnified seventy-five diameters. Externally it resembles dirty green bottle-glass, but shows in the microscope an arborescent crystallization of a green pyroxenic mineral in a colorless felspar base. Plate II, Fig. 48, shows auriferous diorite from Chili, consisting of felspar, with hornblende and crystals of iron pyrites, magnified thirty diameters. Plate II, Fig. 49, is a section of granite from Cornwall, with crystals of orthoclase, hexagonal crystals of brown mica, and colorless quartz, which a higher power shows to contain fluid cavities, magnified twenty-five diameters. Plate II, Fig. 50, a volcanic rock from Peru, composed of felspar, dark crystals of augite, hexagonal crystals of dark mica, and a little magnetic oxide of iron, magnified six diameters. Plate II, Fig. 51, lower silurian roofing-slate, cut at right angles to the cleavage, showing that the latter is not due to crystalline but to mechanical arrangement, magnified two hundred diameters. Plate II, Fig. 52, is an oolitic specimen from Peru, regarded as an eruptive rock by D'Orbigny, but shown in the microscope to be a mere aggregation of sand, etc., without the crystalline character of eruptive rocks.

Materials of Organic Origin.—Rocks and strata derived from plants or animals may be arranged in four groups: 1. The calcareous, or those of which limestones have been formed, as corals, corallines, shells, crinoids, etc. 2. The siliceous, which have contributed to the silica, and may have originated flints, as the microscopic shields of diatoms and siliceous spiculæ of sponges. 3. The phosphatic, as bones, excrement, etc. Fossil excrements are called coprolites, and those of birds in large accumulations, guano. 4. The carbonaceous, or those which have afforded coal and resin, as plants.

To examine the structure of coal, it is necessary to have very thin sections. From its friability, this is a process of great difficulty. The *Micrographic Dictionary* recom-

PLATE II.

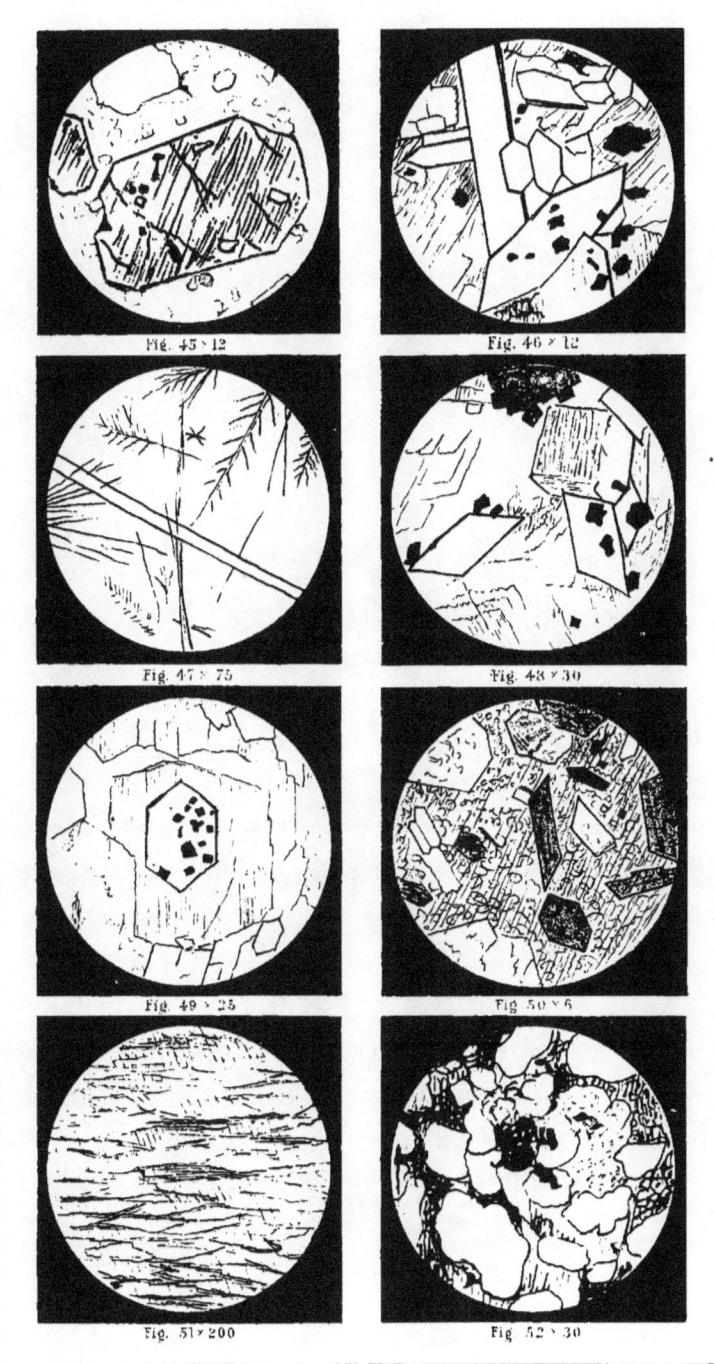

mends the maceration of the coal for about a week in a solution of carbonate of potassium, when thin slices may be cut with a razor. These should be gently heated in nitric acid, and when they turn yellow, washed in cold water and mounted in glycerin, as spirit and balsam render them opaque. Sometimes, as in anthracite, casts of vegetable fibres may be obtained in the ash after burning and mounted in balsam.

The lignites of the tertiary period show a vegetable structure similar to the woods of the present period, but the older coal of the palæozoic series is a mass of decomposed vegetable matter chiefly derived from the decay of coniferous wood, analogous to the araucariæ, as is seen from the peculiar arrangement of the glandular dots on the woody fibres. Traces of ferns, sigillariæ, calamites, etc.. such as are preserved in the shales and sandstones of the coal period, are also met with, but their structure has not been preserved.

Professor Heer, of Zurich, has described and classified several hundred species of fossil plants from the miocene beds of Switzerland by the outlines, nervation, and microscopic structure of the leaves and character of sections of the wood. Several hundred kinds of insects also have been found in the same strata. It is remarkable that a great part of this fossil flora is such as is now common to America, as evergreen oaks, maples, poplars, ternate-leaved pines, and the representatives of the gigantic sequoiæ of California.

The researches of palæontologists have brought to light nearly two thousand species of fossil plants, of which about one-half belong to the carboniferous and one-fourth to the tertiary formations.

The rapid multiplication of the minute microscopic organisms called *diatoms*, is such that Professor Ehrenberg affirms it to have an important influence in blocking up harbors and diminishing the depth of channels. These organisms, now generally regarded as plants, are exceedingly small, and are usually covered by loricæ or shields of pure silica, beautifully marked, as if engraved. These loricæ or shells having accumulated in great quantities, have given rise to very extensive siliceous strata. Thus the "infusorial earth" of Virginia, on which Richmond and Petersburg are built, is such a deposit eighteen feet in thickness. The polishing material called Tripoli, and

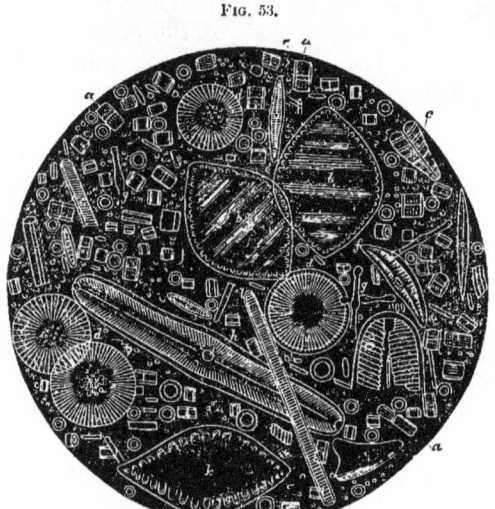

Fig. 53.

Fossil Diatomaceæ, etc., from Mourne Mountain, Ireland: *a, a, a*, Gaillonella (Meloseira) procera, and G. granulata; *d, d, d*, G. biseriata (side view); *b, b*, Surirella plicata; *c*, S, craticula; *k*, S, caledonica; *e*, Gomphonema gracile; *f*, Cocconema fusidium; *g*, Tabellaria vulgaris; *h*, Pinnularia dactylus; *i*, P. nobilis; *l*, Synedra ulna. (From Carpenter.)

the deposit called in Sweden and Norway *berg-mehl* or mountain flour, because used in times of scarcity to mix with flour for bread, are similarly composed. Strata of white rock in the anthracite region of Pennsylvania, and from the sides of the Sierra Nevada and Cascade ranges in California and Oregon, have also been found to consist of such remains (Fig. 53).

The lowest type of animal life, consisting of minute portions of sarcode or animal jelly, having the power of putting forth prolongations of the body at will, contain some forms which cover themselves with shells, usually many-chambered, of carbonate of lime. From the pores in these shells, through which the root-like processes of sarcode are protruded, they are called *Foraminifera*.

Fig. 54.

Fossil Polycystina, etc., from Barbadoes: *a*, Podocyrtis mitra; *b*, Rhabdolithus sceptrum; *c*, Lychnocanium falciferum; *d*, Eucyrtidium tubulus; *e*, Flustrella concentrica; *f*, Lychnocanium lucerna; *g*, Eucyrtidium elegans; *h*, Dictyospyris clathrus; *i*, Eucyrtidium mongolfieri; *k*, Stephanolithis spinescens; *l*, S. nodosa; *m*, Lithocyclia ocellus; *n*, Cephalolithis sylvina; *o*, Podocyrtis cothurnata; *p*, Rhabdolithes pipa. (From Carpenter.)

Another class, the *Polycystina*, secrete a siliceous shell, usually of one chamber. The accumulations of the Foraminifera have formed our chalk beds, while the Polycystina have contributed to siliceous strata, like the *Diatomaceæ* (Fig. 54).

The origin of white chalk strata has been illustrated

by the deep-sea soundings made preparatory to laying the telegraph cable across the Atlantic Ocean. Professor Huxley found the mud composing the floor of the ocean to consist of minute Rhizopods or Foraminifera, of the genus *Globigerina*, together with Polycystina and Diatoms, and a few siliceous spiculæ of sponges. These were connected by a mass of living gelatinous matter, to which he has given the name of *Bathybius*, and which contains minute bodies termed *Coccoliths* and *Coccospheres*, which have also been detected in fossil chalk. It is said that 95 per cent. of the mud of the North Atlantic consists of Globigerina shells.

To examine Foraminifera in chalk, rub a quantity to powder in water with a soft brush, and let it settle for a variable time. The first deposits will contain the larger specimens, with fragments of shell, etc.; the smaller fall next, while the amorphous particles suspended in the water may be cast aside. After drying such specimens as may be selected by the use of a dissecting microscope or Coddington lens, etc., they may be mounted in balsam.

The flint found in chalk often contains Xanthidia, which are the sporangia of Desmidiaceæ, as well as specimens of sponge, Foraminiferal shells, etc. They must be cut as other hard minerals.

There are other deposits besides chalk which are seen by the microscope to consist of minute shells, corals, etc. A section of oolitic stone will often show that each rounded concretion is composed of a series of concentric spheres inclosing a central nucleus which may be a foraminiferal shell. The green sand formation is composed of the casts of the interior of minute shells which have themselves entirely disappeared. The material of these casts, chiefly silex colored with iron, has not only filled the chambers of the shells, but has penetrated the canals of the intermediate skeleton.

The more recent discovery by Drs. Dawson and Carpenter of the organic nature of those serpentine limestones in the Laurentian formations of Canada and elsewhere, which are products of the growth of the gigantic foraminiferal *Eozoon Canadense*, over immense areas of the ancient sea-bottom, is one of still greater interest both to the student of Geology and of Biology.

This immense rhizopod appears to have grown one layer over another, and to have formed reefs of limestone as do the living coral-polyps. Parts of the original skeleton, consisting of carbonate of lime, are still preserved, while certain interspaces have been filled up with serpentine and white augite.

Microscopic Palæontology.—As a general rule it is only the hard parts of animal bodies that have been preserved in a fossil state.

It will often occur that the inspection of a microscopic fragment of such a fossil will reveal with certainty the entire nature of the organism to which it belonged. Thus minute fossil corals, the spines of Echinodermata, the eyes of Trilobites, etc., will determine the position to which we should ascribe the specimen, or a section of tooth or bone will enable the microscopist to assign the fossil to its proper class, order, or family. Thus Professor Owen identified by its fossil tooth, the *Labyrinthodon* of Warwickshire, England, with the remains in the Wittemberg sandstones, and declared it to be a gigantic frog with some resemblances both to a fish, and a crocodile. This prediction the subsequent discovery of the skeleton confirmed.

The minute structure of teeth differs greatly in different animals. In the shark tribe of fishes the dentine is very similar to bone, excepting that the lacunæ of bone are absent. In man and in the Carnivora the enamel is a superficial layer of generally uniform thickness, while in many of the Herbivora the enamel forms with the cementum a series of vertical plates which dip into the substance of the dentine. Enamel is wanting in serpents, Edentata, and Cetacea. Such differences make it quite possible to distinguish the affinities of a fossil specimen from a small fragment of tooth.

In a similar way the microscopic characters of bone vary. The bones of reptiles and fishes have the cancellated structure throughout the shaft, while the lacunæ present very great varieties, so that an animal tribe may be determined by their measurement. In this way many contributions have already been made to palæontology.

CHAPTER VIII.

THE MICROSCOPE IN CHEMISTRY.

THE value of microchemical analysis, and the simplicity of its processes, commend this department of microscopy to general favor.

A large proportion of the actions and changes produced by reagents may be observed as satisfactorily in drops as in larger quantities. The decompositions effected by a galvanic battery far smaller than that contained in a lady's silver thimble, which deflected the mirror at the other end of the Atlantic Telegraph Cable, may be readily observed with a microscope.

Apparatus and Modes of Investigation.—A few flat and hollow glass slides, thin glass covers, test-tubes, small watch-glasses, a spirit-lamp or Bunsen's burner, constitute nearly all the furniture which is essential.

Dr. Wormley[*] directs that a drop of the solution to be examined should be placed in a watch-glass, and a small portion of reagent added with a pipette. The mixture

---

[*] The Microchemistry of Poisons, by Dr. Wormley.

may then be examined with the microscope. If there is no precipitate, let it stand several hours and examine again. Dr. Beale prefers a flat or concave slide, and suggests that if a glass rod be used for carrying the reagent, it must be washed each time, or a portion may be transferred from the slide to the bottle. He also advises the use of small bottles with capillary orifices for reagents. Dr. Lawrence Smith uses small pipettes with the open end covered by india-rubber.

If heat be required, the drop may be boiled on the slide over a spirit-lamp, or a strip of platinum-foil or mica may be held with forceps so as to get a red or white heat from the lamp or a Bunsen burner. This is especially needed to get rid of organic matters.

For the examination of earthy materials, as carbonate or phosphate of lime, phosphate of ammonia and magnesia, sulphates or chlorides, a small fragment may be placed on a slide and covered with thin glass. A drop of nitric acid is then put near the edge of the cover. If bubbles escape a carbonate is indicated. Neutralize the acid with ammonia; let the flocculent precipitate stand awhile; cover and examine with the microscope. After a time, amorphous granules and prisms will show phosphates of ammonia, magnesia, and lime. Sulphates are shown by adding to the nitric acid solution nitrate of barytes, and chlorides by nitrate of silver.

Dr. Beale recommends adding glycerin to the test solutions. The reactions are slower but more perfect, and the crystalline forms resulting are more complete.

If a sublimate be desired, a watch-glass can be inverted over another, and the lower one containing the material, as biniodide of mercury, etc., heated over a spirit-lamp, or the sublimation may be made in a reduction-tube.

Preparation of Crystals for the Polariscope.—Many specimens may be prepared by concentrating the solution with heat and allowing it to cool. It should not be evaporated to dryness. Many salts may be preserved in balsam, but some are injured by it, and need glycerin or castor oil as a preserving fluid.

The method of crystallization may be modified in various ways so as to obtain special results. Thus if a solution of sulphate of iron is suffered to dry on a slide, the crystals will be arborescent and fern-like, but if the liquid is stirred with a glass rod or needle while evaporating, separate rhombic prisms will form, which give beautiful colors in the polariscope. Pyrogallic acid also crystallizes in long needles, but a little dust, etc., as a nucleus, brings about a change of arrangement resembling the "eye" of the peacock's tail.

A saturated solution dropped into alcohol, if the salt is insoluble in alcohol, will produce instantaneous crystals.

To obtain the best results, some crystals, as salicin, should be fused on a slide over the lamp, and the matter spread evenly over the surface. This may be done with a hot needle. The temperature greatly affects the character of the crystallization. If very hot, the crystals run in lines from a common centre. A medium temperature produces concentric waves.

Many new forms result from uniting different salts in different proportions. The knowledge of these different effects can only be attained by experience.

Sections of crystals, as nitrate of potash, etc., to show the rings and cross in the polariscope, are difficult to make. After cutting a plate with a knife to about one-fourth of an inch thick, it may be filed with a wet file to one-sixth of an inch, smoothed on wet glass with fine emery, and polished on silk strained over a piece of glass, and rubbed with a mixture of rouge and tallow. The nitre must be rubbed till quite dry, and the vapor of the fingers prevented by the use of gloves.

For a general account of the use of polarized light, see Chapter VI.

The Use of the Microspectroscope.—We have already described this accessory in Chapter III. It promises important results in chemical analysis, but requires delicate observation and exact measurements, together with a careful and systematic study of a large number of colored substances.

In using the microspectroscope, much depends on the regulation of the slit. It should be just wide enough to give a clear spectrum without irregular shading. As a general rule, it should be just wide enough to show Fraunhofer's lines indistinctly in daylight. The slit in the side stage should be such that the two spectra are of equal brilliancy. No light should pass up the microscope but such as has passed through the object under examination. This sometimes requires a cap over the object-glass, perforated with an opening of about one-sixteenth of an inch for a one and a half inch objective.

The number, position, width, and intensity of the absorption-bands are the data on which to form an opinion as to the nature of the object observed, and Mr. Sorby has invented a set of symbols for recording such observations. (See Dr. Beale's *How to Work with the Microscope.*) These bands, however, do not relate so much to the elementary constitution as to the physical condition of the substance, and vary according to the nature of the solvent, etc., yet many structures give such positive effects as to enable us to decide with confidence what they are.

Colored beads obtained by ordinary blowpipe testing, sections of crystals, etc., cut wedge-shaped so as to vary their thickness, often give satisfactory results. But minute quantities of animal and vegetable substances, as blood-stains, etc., dissolved and placed in short tubes fastened endwise on glass slides, or in some other convenient apparatus, offer the most valuable objects of research.

To measure the exact position of the absorption-bands,

the micrometer already described may be used, or Mr. Sorby's apparatus, giving an interference spectrum with twelve divisions, made by two Nicol's prisms, with an intervening plate of quartz of the required thickness.

The value of this mode of investigation in medical chemistry, and for purposes of diagnosis or jurisprudence, may be seen by the following illustrations:*

Pettenkofer's Test for Bile (Fig. 55).—To a few drops of bile in a porcelain dish, add a drop of solution of cane-

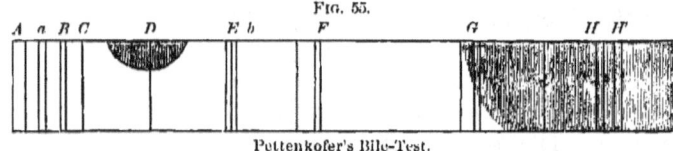

Pettenkofer's Bile-Test.

sugar, and then concentrated sulphuric acid drop by drop, with agitation. The mixture becomes a purple-red color, and shows a spectrum as in the figure. The color will be destroyed by water and alcohol.

Tests for Blood.—Hæmatocrystalline, or cruorin, composed of an albuminoid substance and hæmatin, generally crystallizes in tetrahedra or octahedra. In blood from

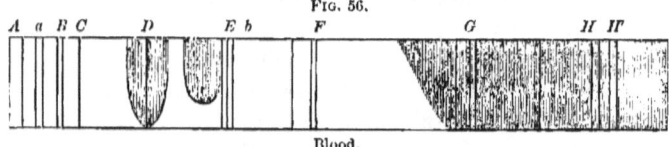

Blood.

the horse and from man only an amorphous deposit is found. The watery solution of this substance properly diluted, shows two remarkable bands of absorption, and obscuration of the blue and violet end of the spectrum (Fig. 56). As the blood of all vertebrates shows the same

---

* See Thudichum's Manual of Chemical Physiology. New York, 1872.

bands, it is judged that hæmatocrystalline is present in it as such, and not formed from it. By treating a solution of blood which exhibits the two absorption-bands with hydrogen, or with a solution of ferrous sulphate containing tartaric acid and excess of ammonia, taking care to

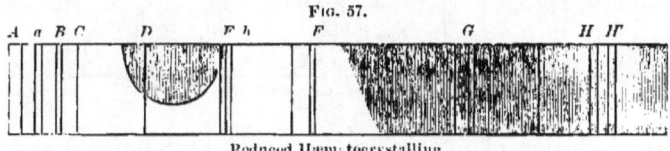

Reduced Hæmatocrystalline.

exclude the air, the color of the solution changes to purple, and the spectroscope shows only one broad band instead of two (Fig 57). Shaking with air will restore the two bands. By treating blood with hydrothion or am-

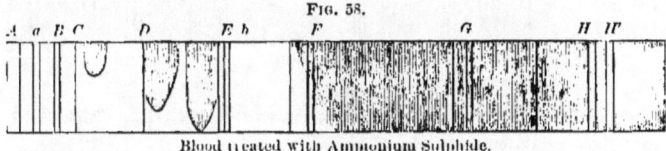

Blood treated with Ammonium Sulphide.

monium sulphide, three bands make their appearance, as in Fig. 58.

Hæmatin is seen by the microscope to consist of small

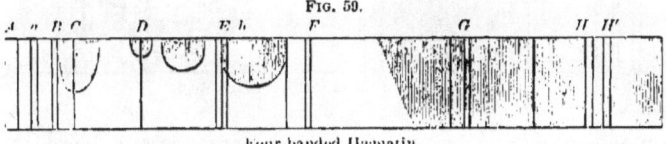

Four-banded Hæmatin.

rhombic crystals. Dissolved in alcohol and a little sulphuric acid, the spectrum shows four, and under some circumstances five, bands (Fig. 59). Rendered alkaline by caustic potash, one broad band appears (Fig. 60). Acid will restore the former spectrum.

Dissolve hæmatin in water with a little caustic potash.

Alkaline Hæmatin.

To a solution of ferrous sulphate, add tartaric acid and then ammonia till alkaline. Pour a little of the clear mixture into the hæmatin solution. The spectrum of re-

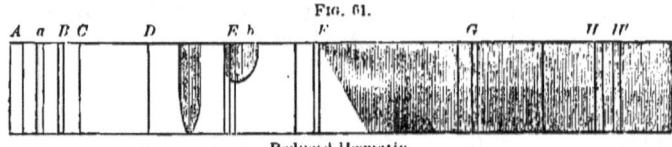

Reduced Hæmatin.

duced hæmatin will show two bands (Fig. 61). Shaking with air will restore the former spectrum.

Lutein Spectra.—The juice of the corpora lutea, to which sulphuric acid and a little sugar is added, gives a fine

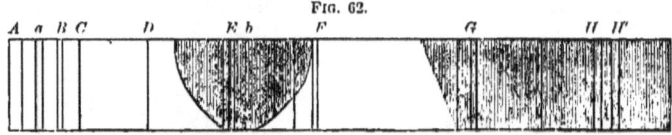

Juice of Corpora Lutea with Sulphuric Acid.

purple color, and shows in the spectroscope one band in the green (Fig. 62). Its chloroform solution, examined with lime-light, shows two bands in blue (Fig. 63). An alcoholic or ethereal solution gives a third one in the violet.

Cysto lutein, or the yellow fluid of an ovarian cyst, shows with the lime-light three bands in blue, in the

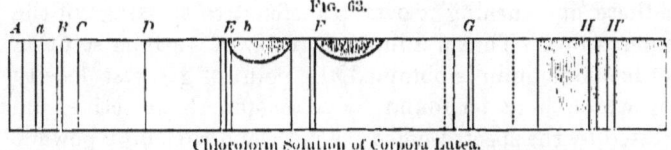

Chloroform Solution of Corpora Lutea.

same position as the chloroform solution of lutein (Fig. 64).

The serum of blood, etc., shows the bands of hæmato-

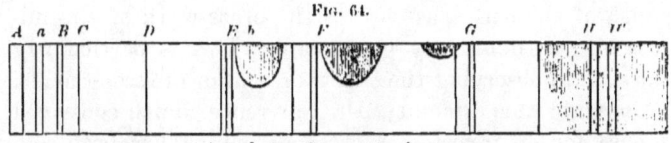

Cysto-Lutein from an Ovarian Cyst.

crystalline and one or two doubtful bands, as in the figure (Fig. 65).

Dr. Richardson, of Philadelphia, gives the following directions for examining blood-stains: Procure a glass slide with a circular excavation, and moisten the edges of the cavity with a small drop of diluted glycerin. Lay

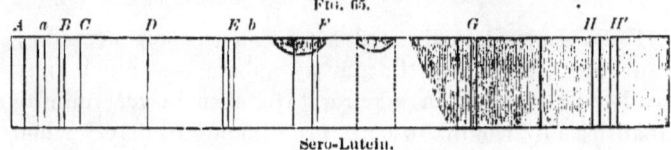

Sero-Lutein.

a clean glass cover, a little larger than the excavation, on white paper, and put on it the smallest visible fragment of blood-clot. With a needle, put on the centre of the cover a speck of glycerin, not larger than a full stop (.), and with a dry needle push the blood to the edge that it may be just moistened with the glycerin. Place the slide on the cover so that the glycerin edges of the cavity may adhere, and turning it over, transfer it to the stage of the microscope. Thus a minute quantity of a strong solution of hæmoglobulin is obtained, the point of greatest density of which may be found by a one-fourth objective, and tested by the spectroscopic eye-piece and with high powers. The tiny drop may be afterwards wiped off with moist blotting-paper, and a little fresh tincture of guaiacum added, showing the blue color of the guaiacum blood-test.

Inverted Microscope of Dr. Lawrence Smith.—In ordinary chemical investigations there is some risk of injuring the polish of the lenses, as well as the brass work of the microscope, without very great care. This is particularly the case in observing the effects of heat or of strong acids. To obviate this difficulty, Dr. Lawrence Smith contrived a plan for an inverted microscope, which has been constructed by Nachet of Paris. The optical part of the instrument is below the stage, and is furnished with a peculiar prism, by which the rays from the objective are bent into a conveniently inclined body. The illuminating apparatus is above the stage. This construction renders the instrument well adapted to chemical investigations.

General Microchemical Tests.

Dr. Wormley has directed attention to some necessary cautions. He shows that many substances which may readily be detected in a pure state, even in very minute quantities by the microscope, are difficult to detect when mixed with complex organic materials. This is especially applicable to the alkaloids, which should be separated from such mixtures by the use of the dialyzer—a hoop with a bottom of parchment-paper, etc.—or extracted with ether or chloroform.

The purity of all reagents should be carefully established, and they should be kept in hard German glass bottles, and only distilled water used in all our researches.

The true nature of a reaction that is common to several substances may often be determined with the microscope. Thus a solution of nitrate of silver becomes covered with a white film when exposed to several different vapors, but hydrocyanic acid is the only one which is crystalline. This will detect 100,000th of a grain of the acid. A slip of clean copper boiled in a hydrochloric acid solution of arsenic, mercury, antimony, etc., becomes coated with the metal, but when heated in a reduction-tube, arsenic only yields a sublimate of octahedral crystals, and mercury only will furnish metallic globules.

A solution of iodine produces distinct reaction with 100,000th of a grain of strychnine in solution in 1 grain of water, but as this is common to other alkaloids, other tests are needed. Yet the absence of such a reaction shows the absence of the alkaloid.

The degree of dilution is important. Thus bromine with atropin yields a crystalline deposit from 1 grain of a 20,000th or stronger dilution, but not with diluter solutions. A limited quantity of sulphuretted hydrogen throws down from corrosive sublimate a white deposit, while excess produces a black precipitate.

Blue and Reddened Litmus Paper are used as tests for acids and alkalies. It is a bibulous paper dyed in infusion of litmus. The red is made by adding a little acetic acid to the infusion. Dry substances and vapors require the paper to be moistened with distilled water. If the acid reaction depends on carbonic acid, warming the paper on a slide over a lamp will restore the color. So if a volatile alkali, ammonia or carbonate of ammonia, have made the red paper blue, its color will be restored by a gentle heat. Sometimes the infusion of litmus is more convenient than the paper.

Alcohol coagulates albuminous matter.

Ether dissolves fat.

Acetic Acid will dissolve phosphate or carbonate of lime, but not the oxalate.

Nitrate of Barytes in cold saturated solution is a test for sulphuric and phosphoric acids. The precipitated sulphate of baryta is insoluble in acids and alkalies. The phosphate is soluble in acids and insoluble in ammonia.

Nitrate of Silver.—A solution of 60 grains to the ounce of water is a convenient test for chlorides and phosphates. Chloride of silver is white, soluble in ammonia and insoluble in nitric acid. The tribasic phosphate of silver is yellow, and soluble in excess of ammonia or of nitric acid.

Oxalate of Ammonia is a test for salts of lime. Dissolve the material in nitric acid, and add excess of ammonia. Dissolve the flocculent precipitate in excess of acetic acid, and add the oxalate of ammonia. Oxalate of lime is insoluble in alkalies and acetic acid, but soluble in strong mineral acids.

Iodine is a test for starch, coloring it blue. Albuminous tissues are colored yellow, and vegetable cellulose a brownish-yellow. The addition of sulphuric acid turns cellulose blue.

Determination of Substances.

ALKALIES.

Bichloride of platinum precipitates from salts of potash or ammonia a yellow double chloride, which crystallizes in beautiful octahedra. It has no precipitating effect on solutions of soda. Polarized light will distinguish the 800,000th of a grain of double chloride of sodium and platinum by its beautiful colors from the chloride of potassium and platinum, or of platinum alone. The double chloride of platinum and potassium may be distinguished from that of ammonia by heating to redness, treating with hot water, and acting on with nitrate of silver. The ammonium compound after ignition leaves only the platinum, which gives no precipitate with nitrate of silver, while the potassium chloride yields a white precipitate of chloride of silver.

Antimoniate of potash throws down from solutions of soda and its neutral salts a white crystalline antimoniate of soda, the forms of which vary according to the strength of the solution; generally they are rectangular plates and octahedra.

ACIDS.

Sulphuric.—In solutions acidulated with hydrochloric or nitric acid, the chloride or nitrate of baryta produces a white precipitate. Veratrin added to a drop of concentrated sulphuric acid produces a crimson solution, or deposit if evaporated.

Nitric.—Heated with excess of hydrochloric acid eliminates chlorine, which will dissolve gold leaf. A blood-red color is produced when nitric acid or a nitrate is mixed with a sulphuric acid solution of brucin.

Hydrochloric.—Nitrate of silver precipitates amorphous chloride of silver; soluble in ammonia, but insoluble in nitric and sulphuric acid.

Oxalic.—Nitrate of silver precipitates amorphous oxalate of silver; soluble in nitric acid and also in solution of ammonia.

Hydrocyanic.—Put a drop of acid solution in a watch-glass, invert another over it containing a drop of solution of nitrate of silver, and a crystalline film will form. A solution of hydrocyanic acid treated with caustic potash or soda and then with persulphate of iron yields Prussian blue.

Phosphoric.—A mixture of sulphate of magnesia, chloride of ammonium, and free ammonia produces in solutions of free phosphoric acid and alkaline phosphates white feathery or stellate crystalline precipitate of ammonio-phosphate of magnesia. A slower crystallization gives prisms.

METALLIC OXIDES.

These may usually be determined by treating a small portion of solution, acidulated with hydrochloric acid, by sulphuretted hydrogen; another, and neutral portion with sulphuret of ammonium; and a third with carbonate of soda.

Antimony.—Sulphuretted hydrogen throws down orange-red precipitate from tartar-emetic solutions, etc.

Arsenic yields white octahedral crystals of arsenious acid when sublimed. Arsenious acid may be reduced to metallic arsenic by heating to redness in a tube with charcoal and carbonate of soda. A solution of arsenious acid yields octahedral crystals by evaporation, so as to determine with the microscope 1000th to 10,000th of a grain.

Ammonio-nitrate of silver throws down from an aqueous solution of arsenious acid a bright yellow precipitate, ammonio-sulphate of copper a green precipitate, and sulphuretted hydrogen a bright yellow.

Mercury.—Bichloride of mercury, moistened with a drop of solution of iodide of potassium, assumes the bright scarlet color of biniodide of mercury. A strong solution of caustic potash or soda turns bichloride of mercury yellow from the formation of protoxide; but calomel or chloride of mercury is blackened from formation of suboxide. Heated in a reduction-tube with dry carbonate of soda, the sublimate shows under the microscope small, opaque, spherical globules of mercury. Dr. Wormley states that a globule of mercury or "artificial star" may be discriminated by the one-eighth objective if it be but the 25,000th of an inch in diameter, weighing about the 9,000,000,000th of a grain; globules of $\frac{1}{5000}$th of an inch diameter weigh about 70,000,000th of a grain.

Lead.—Sulphuretted hydrogen gives a black amorphous deposit. Sulphuric and hydrochloric acids yield a white precipitate. Chloride of lead crystallizes in needles. Iodide of potassium gives a bright yellow precipitate, soluble in boiling water, and crystallizing in six-sided plates. Bichromate of potassium yields a bright yellow amorphous deposit.

Copper.—Sulphuretted hydrogen gives a brown or blackish deposit; ammonia a blue or greenish-blue amorphous precipitate, or in dilute solutions a blue color to the liquid; caustic alkali, a similar precipitate, which on boiling in excess of reagent becomes black, but if grape-sugar, or some other organic agents, be present, a yellow or red precipitate of suboxide of copper occurs. Arsenite of potassium produces a bright green.

Zinc.—Sulphuretted hydrogen gives a white amorphous deposit—the only white sulphuret. Alkalies produce a white hydrated oxide of zinc.

ALKALOIDS.

The editors of the *Micrographic Dictionary* refer to a paper of Dr. T. Anderson, in the *Edinburgh Monthly Journal*, where he shows that the microscope readily distinguishes the more common alkaloids from each other by the form of their crystals and of their sulphocyanides. The alkaloids are first dissolved in dilute hydrochloric acid, then precipitated on a glass plate with a solution of ammonia, or if the sulphocyanide is required, with a strong solution of sulphocyanide of potassium. It may then be placed under the microscope. The solution should not be too concentrated. This branch of investigation has been greatly promoted by the elegant work of Dr. Wormley, already referred to, on the *Microchemistry of Poisons*.

Atropin.—Ammonia throws down an amorphous precipitate. One grain of a $\frac{1}{100}$th grain solution yields to caustic potash or soda a precipitate which, when stirred with a glass rod, becomes a mass of crystals, as in Plate III, Fig. 66. The sulphocyanide of potassium gives no precipitate.

Aconitin.—No characteristic test, except the physiological one; $\frac{1}{1000}$th of a grain produces on the end of the tongue a peculiar tingling and numbness, lasting for an hour; $\frac{1}{100}$th grain in alcohol, rubbed on the skin, produces temporary loss of feeling.

Brucin or Brucia.—Potash or ammonia produces stellar crystals. Sulphocyanide of potassium, feathery, or sheaf-like. (Plate III, Fig. 67.) Nitric acid produces a blood-red color, changing to yellow by heat. On cooling the latter and adding protochloride of tin, it becomes a beautiful purple. Ferricyanide of potassium, with $\frac{1}{100}$th grain of brucin yields the most brilliant polariscope crystals. (Plate III, Fig. 68).

Cinchonine.—Ammonia produces granular radiating crystals. (Plate III, Fig. 69.) Sulphocyanide of potassium six-sided plates, some irregular. (Plate III, Fig. 70.)

Conine.—This alkaloid and nicotin are distinguished from other alkaloids by being liquid at ordinary temperatures, and by their peculiar odor. Conine may be known from nicotin by its odor and sparing solubility in water, by yielding crystalline needles to the vapor or solution of hydrochloric acid, a white precipitate with corrosive sublimate, and a dark-brown precipitate with nitrate of silver.

Codein.—Ammonia or alkalies give a white amorphous deposit. Sulphocyanide of potassium, crystalline needles. A solution of iodine in iodide of potassium, a reddish-brown precipitate, which becomes crystalline. This is soluble in alcohol, from which it separates in plates (Plate III, Fig. 71), which appear beautiful in the polariscope.

Daturin.—According to Dr. Wormley, this is identical with atropin.

Narcotin.—In its pure state crystallizes in rhombic

PLATE III.

Fig. 66.

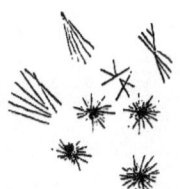

Fig. 67.

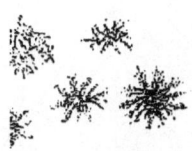

Fig. 68.

Fig. 69.

Fig. 70.

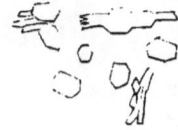

Fig. 71.

Fig. 72.

Fig. 73.

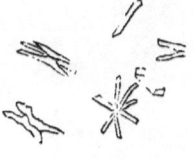

Fig. 74.

Fig. 75.

prisms, or oblong plates. Ammonia, the alkalies, and their carbonates produce tufts of crystals (Plate III, Fig. 72). A drop of aqueous solution of a salt of narcotin, exposed to vapor of ammonia, is covered with a crystalline film if it only contains $\frac{1}{5000}$th of its weight of alkaloid.

Morphine.—When pure crystallizes in short rectangular prisms. Sulphuric acid dissolves them, and if bichromate of potash be added, green oxide of chromium results. Concentrated nitric acid turns it orange-red, and dissolves it. A strong solution treated with a strong solution of nitrate of silver and gently heated, decomposes the latter and produces a shining crystalline precipitate of metallic silver. In dilute solutions, alkalies precipitate a crystalline form (Plate III, Fig. 73). No precipitate with sulphocyanide of potassium unless highly concentrated.

Quinine.—Amorphous precipitate with ammonia. Sulphocyanide of potassium gives irregular groups of acicular crystals, like those produced by strychnine, but longer and more irregular (Plate III, Fig 74). The solution should be dilute, and twenty-four hours allowed for the crystals to form.

The iodo-disulphate, or Herapathite, gives crystals of a pale olive-green color, which possess a more intense polarizing power than any other known substance. Dr. Herapath proposed this as a delicate test for quinine. A drop of test-liquid—made with 3 drachms of acetic acid, 1 drachm of rectified spirits, and 6 drops of dilute sulphuric acid—is placed on a slide and the alkaloid added. When dissolved a little tincture of iodine is added, and after a time the salt separates in little rosettes. By careful manipulation crystals of this salt may be formed large enough to replace Nicol's prisms or tourmaline plates in the polarizing apparatus. When the crystals of Herapathite cross each other at a right-angle, complete blackness results. Intermediate positions give a beautiful play of colors.

Strychnine.—Ammonia gives small prismatic crystals,

some crossed at 60° (Plate III, Fig. 75). Sulphocyanide of potassium produces flat needles, often in groups. Iodine in iodide of potassium gives a reddish-brown amorphous precipitate, crystalline in dilute solutions. When pure, strychnine appears in colorless octahedra, lengthened prisms or granules. To a solution of the alkaloid or its salts in a drop of pure sulphuric acid, which produces no color, add a small crystal of bichromate of potash, and stir slowly with a pointed glass rod. A blue color will appear, passing into purple, violet, and red. The bright yellow crystals of chromate of strychnia, if dried and touched with sulphuric acid, will also show the color test. This is said to be delicate enough to show $\frac{1}{100,000}$th of a grain of strychnine. The tetanic convulsions of frogs immersed in a solution of strychnine, or after injections of the solution in lungs or stomach, etc., is also a very delicate test.

Veratrin and its salts treated in the dry state with concentrated sulphuric acid, slowly dissolve to a reddish-yellow, or pink solution, which becomes crimson-red. The process is accelerated by heat.

Narcein, touched with the cold acid, becomes brown, brownish-yellow, and greenish-yellow, and if heated, a dark purple-red.

Solanin turns orange-brown, and later purplish-brown.

Piperin turns orange-red to brown.

Salicin gives to the acid a crimson pink, changing to black.

Papaverin gives a fading purple.

CRYSTALLINE FORMS OF VARIOUS SALTS.

Our limits forbid extended description, yet a few forms of frequent recurrence will be useful to the student. For crystals in plants or from animal secretions reference may be made also to succeeding chapters.

Salts of Lime.—The *carbonate* sometimes occurs in ani-

Fig. 76.

Fig. 77.

Fig. 78.

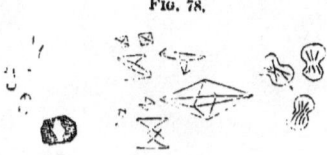

Fig. 79.

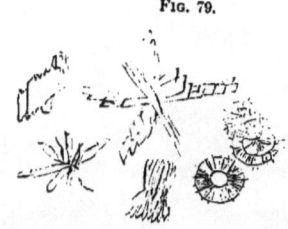

Fig. 80.

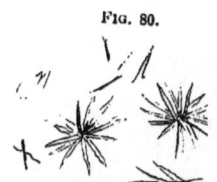

Fig. 81.

Fig. 82.

Fig. 83.

Fig. 84.

mal secretions in the form of little spheres or disks, consisting of groups of radiating needles. In otoliths it is often in minute hexagonal prisms with trilateral summits. It is deposited from water in irregular forms, all of which are grouped needles. Sometimes it assumes the rhombohedral form, as in the oyster shell (Plate IV, Fig. 76). In any doubtful case, test as described at pages 99 and 108.

Lactate of Lime gives microscopic crystals, consisting of delicate radiating needles (Plate IV, Fig. 77).

Oxalate of Lime occurs as square flattened octahedra, as square prisms with quadrilateral pyramids, as fine needles, and as ellipsoidal flattened forms, sometimes constricted so as to resemble dumb-bells (Plate IV, Fig. 78).

Phosphate of Lime is usually in the form of thin rhombic plates (Plate IV, Fig. 79).

Sulphate of Lime rapidly formed, as in chemical testing, gives minute needles or prisms (Plate IV, Fig. 80). When more slowly formed, these are larger and mixed with rhombic plates.

Soda Salts.—*Chloride of Sodium* or common salt generally forms a cube, terminated by quadrangular pyramids or depressions (Plate IV, Fig. 81). The crystals do not polarize light.

Plate IV, Fig. 82, represents crystals of *oxalate* of soda, and Plate IV, Fig. 83, those of *nitrate*.

Magnesia Salts.—*Ammonio-phosphate*, or triple phosphate, is often found in animal secretions. The most common form is prismatic, but sometimes it is feathery or stellate (Plate IV, Fig. 84).

Sulphate of Magnesia forms an interesting polarizing object.

A most instructive series of salts may be made by rapidly crystallizing some on glass slides, and allowing others to deposit more slowly. In this way a set of specimens may be prepared for comparison.

CHAPTER IX.

THE MICROSCOPE IN BIOLOGY.

The science of biology (from βιος, life), which treats of the forms and functions of living beings, would be crude and imperfect without the aid of the microscope. Whatever might be learned by general observation, we should miss the fundamental laws of structure and the unity which we now know pervades distant and apparently different organs, as well as distinct species, if we were deprived of the education which microscopy gives the eye and hand.

The evident differences between living and non-living bodies led to ancient theories of life which are still influential in modern thought, but neither microscope nor scalpel nor laboratory have revealed the mystery which seems ever to beckon us onward to another and entirely different sphere of existence. Hippocrates invented the hypothesis of a principle (φυσις, or nature) which influences the organism and superintends it with a kind of intelligence, and to which other principles (δυναμεις, powers) are subordinated for the maintenance of various functions. This was also the theory of Aristotle, who gave the name of soul (ψυχη) to the animating principle.

Paracelsus and the chemical philosophers, from the fifteenth to the seventeenth century, maintained that all the phenomena of vitality may be explained by chemical laws. To these succeeded the mathematical school under Bellini (A.D. 1645), who taught that all vital functions may be explained by gravity and mechanical impulse. These theories were supplanted by those of the physiologists. Van Helmont revived the Hippocratian idea of a specific agent, which he called *archeus*. This was more fully elaborated by Stahl, who taught that by the operation of an immaterial animating principle or soul (*anima*), all vital functions are produced. The *vis medicatrix naturæ* of Cullen was an attempt to compromise between the rival theories of a superadded principle and a special activity in organized matter itself.*

Harvey, Hunter, Müller, and Prout proposed hypotheses similar to those of Aristotle and Hippocrates, and many modern scientific men accept similar views. The recent doctrine of the correlation of physical forces has, however, revived the mechanical and chemical theories, and the industry with which these views have been propagated has gained many adherents.

It is to be regretted that philosophy should assume the name of science and dogmatize under that appellation. The object of science is to state facts, and not to dream, yet such is the nature of man's intellect that it will seek to account for facts, and is thus drawn into metaphysical speculation. If the age-long controversy between the physicists and the vitalists is ever to cease, it will probably be through the microscopic demonstration of the absolute difference between living and non-living matter.

In the present chapter it is designed to set forth briefly the principal facts of elementary biology as they have been brought to light by microscopy. For further illustrations in vegetable and animal histology, reference may be made to following chapters.

1. All biologists agree that *the elementary unit in living bodies* is the cell. This, according to the most recent investigations, is a soft, transparent, colorless, jelly-like particle of matter, which may be large enough to be just discernible to the naked eye, or so small as to be invisible with our best instruments. The simplest or most elementary forms of vegetable or animal life consist of single cells, while the more complex organisms are built up of

---

* Compare Bostock's History of Medicine.

great numbers of these cells with the materials which they have produced and deposited.

Haller, who has been called the father of modern physiology, seems first to have conceived, though vaguely (A.D. 1766), the idea of the essential unity of vital structure.

In 1838, Schleiden and Schwann wrote on the elementary cell, the former treating of the vegetable, and the latter of the animal cell. From this time may be dated the origin of the cell doctrine. Much importance was assigned to the distinction between cell-wall, cell-contents, nuclei, and nucleoli.

In 1835, Dujardin discovered in the lower animals a contractile substance capable of movement, to which he gave the name of *sarcode*.

In 1861, Max Schultze showed that sarcode is analogous to the body or contents of animal cells, and that on this account the infusorial animalcules possessed of independent life were simple or compound.

Examinations of this structure were made by numerous observers, and the identity of many of its properties in animals and vegetables established. To this structure the name of *protoplasm*, rather than sarcode, has been assigned. As this term has been somewhat loosely used, so as to refer to it either in the dead or living state, Dr. Beale has proposed the term *bioplasm* for elementary structure while living, and has given a generalization from observed facts which has attracted much attention. He distinguishes in all organic forms three states of matter: First. Germinal matter or *bioplasm*, or matter which is living. Second. Matter which was living, or *formed material*. Third. Matter about to become living, or *pabulum*.

Schleiden and Schwann considered the cell as a growth from a nucleus, and to consist of a cell-wall and cavity. In vegetable cells there seemed to be an external wall of cellulose, within which was another, the primordial utricle. But it has since been shown that the appearance of the primordial utricle is caused by the protoplasm or bioplasm lying in apposition with the inner surface of the cell-wall. In the cryptogamia, cells are known to occur in which no nucleus is visible. Max Schultze and Häckel have also discovered non-nucleated forms of animal life. The idea of nucleus and cell-wall as essential to a cell is therefore abandoned. Nuclei are regarded as new centres of living matter, or minute particles of such matter capable of independent existence. Some of these masses are so small as to be barely visible with the one-fiftieth objective under a magnifying power of five thousand diameters.

2. *The structure and formation of a simple cell* may be illustrated by Plate V, Figs. 85 to 89, after Beale.* The earliest condition of such a living particle is shown in Plate V, Fig. 85. If the external membrane of a fully developed spore or any of the growing branches (Plate V, Figs. 86 to 89) be ruptured, such particles would be set free in vast numbers.

The surface of such a particle becomes altered by contact with external agencies. A thin layer of the external surface is changed into a soft membrane or cell-wall, through which pabulum passes and undergoes conversion into living matter, which thus increases. The increase of size is not owing to the addition of new matter upon the external surface, but to the access of new matter interiorly. The thickness of the formed material depends on external circumstances, as temperature, moisture, etc. If these be unfavorable to the access of pabulum, layer after layer of living matter will die or be deposited, as in Plate V, Figs. 87 and 88. If such a cell be exposed to circumstances favorable to growth, the accession of fresh pabulum will cause portions of living matter to make

---

* Physiological Anatomy and Physiology of Man, by Drs. Todd, Bowman, and Beale. New edition.

their way through natural pores or chance fissures and protrude, as in the figures.

3. *The peculiar phenomena of living cells* or bioplasms may be classified as follows: Active or spontaneous movement, nutrition and growth, and the power of reproduction. These vital actions, according to Dr. Beale, occur in the bioplasm only, while the formed material, or non-living matter, is the seat of physical and chemical changes exclusively. Physical processes, as diffusion and osmose, occur in bioplasmic particles, but the peculiar phenomena referred to, and which are properly termed vital, do not occur in non-living matter.

Movements of Cells.—Granules imbedded in the bioplasm, either formed material or accidental products, enable our microscopes to observe internal movement, while change of form and of place exhibit the movement of the entire cell.

The granular movement is either vibratory or continuous. The vibrations of the granules appear similar to the molecular movement described by Dr. Robert Brown in 1827, and which is common to all small masses of matter, organic or inorganic. Minute cells may thus dance in fluid as well as fine powders, etc. Such movements occur, however, in the interior of living cells, and may possibly be connected with vitality. In the salivary corpuscles, the dancing motion ceases on the addition of a solution of one-half to one per cent. of common salt, while such addition has no influence of the kind on fresh pus or lymph.

The continuous granular motion is either a relatively slow progression, corresponding to the change of form in the cell, or a swifter flowing movement. Max Schultze thus describes this motion in the threads of sarcode projected from the apertures of a Foraminiferal shell: "As the passengers in a broad street swarm together, so do the granules in one of the broader threads make their

PLATE V.

Fig. 85.

Minute particles of Bioplasm. From Mildew. $\frac{1}{50}$th in. Obj.

Fig. 88.

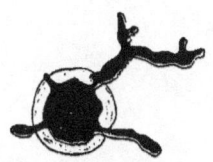

Passage of Germinal-matter through pores in the formed material. × 1800.

Fig. 86.

Production of formed-material on surface of Bioplasm. × 1800.

Fig. 89.

Production and accumulation of Formed-material on Bioplasm. Epithelium of cuticle. × 700.

Fig. 87.

 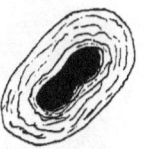

Further production of formed-material. At a is the budding Bioplasm, passing through pores in the formed-material. × 1800.

Fig. 90.

Amœbæ from organic infusion.

way by one another, oftentimes stopping and hesitating, yet always pursuing a determinate direction corresponding to the long axis of the thread. They frequently become stationary in the middle of their course, and then turn round, but the greater number pass to the extreme end of the thread, and then reverse the direction of their movement." No physical or chemical action with which we are acquainted will account for such motions, which have no analogy in unorganized bodies.

Changes of form are most strongly marked in the lower forms of animal life, although occurring also in the simpler vegetables, as the volvox. The *Amœba* or Proteus is typical of such changes, which have hence been termed *Amœboid* (Plate V, Fig. 90). When an Amœba meets another animal which is too slow to escape, it sends out projections which encircle its prey; these coalesce, and invest the whole mass with its bioplasm. It maintains its grasp till it has abstracted all the portions which are soluble, and then relaxes its hold.

Amœboid cells in higher animals rarely move so rapidly as the Amœba itself. Their motions are limited to a gradual change of form or to the protrusion of processes in the form of threads, or tuberosities, or tufts, which either drag the rest of the body after them or are again withdrawn.

Cells of bioplasm may not only change their form, but may wander from place to place by protruding a portion of their mass, which drags the rest after it. The discovery of wandering cells in the higher organisms, as man, has opened quite a new and important field of physiological and pathological research.

The movements of bioplasm may be changed, accelerated, retarded, or stopped by a variety of stimuli, mechanical, electrical, chemical, and nervous. Gentle warmth and moisture are necessary to their perfection.*

---

* See Stricker's Manual of Histology.

The nutrition and growth of the living cell has already been described as the conversion of pabulum into bioplasm or living matter. The subject of reproduction will be examined below under the head of *cell-genesis*.

4. *The microscopic demonstration of bioplasm* may be effected by the use of an alkaline solution of coloring matter, as carmine. (See Chapter V.) As bioplasm possesses an acid reaction, the alkali is neutralized and the color retained. This process, however, is rather a demonstration of the protoplasm which was recently alive. For living cells or bioplasm, we must depend on supplying them artificially with colored food. Thus indigo, carmine, etc., in fine particles, added to the pabulum of cells or liquid media in which they float, will be taken into the interior of the bioplasm by the nutritive process. In this way Recklinghausen showed the migration of pus-corpuscles.

Welcker and Osborne were the first to use a solution of carmine in order to stain the nuclei of tissues. They were followed by Gerlach and Beale, the latter of whom has greatly improved the process and shown its significance.

5. *The chemistry of cells and their products* is an essential part of biology, but would lead us too far from our subject to discuss, yet a few points may not be irrelevant.

The chemical composition of bioplasm consists essentially of oxygen, hydrogen, nitrogen, and carbon. Other elements are often present and important, but not essential. Of the relation of these elements we know nothing, save that they are in a state of constant vibration or change. Dr. Beale considers it doubtful if ordinary chemical combination is possible while the matter lives. Analysis in the laboratory is only possible with the compounds resulting from the death of the cell.

When living or germinal matter is converted into formed material, a combination of its elements takes place, often with very complex results, the nature of which has hitherto baffled the efforts of chemists to determine. When the life of germinal matter, however, is suddenly destroyed, or rather when the matter is first transformed, the compounds resulting from various species have similar chemical composition and properties, and an acid reaction is developed. Fibrin, albumen, water, and certain salts may thus be obtained from every kind of germinal matter. Fatty matters also result, which continue to increase in quantity for some time after death. In slow molecular death, a certain amount of oxygen is taken into combination, which gives rise to different results from those which occur when life is suddenly destroyed. Still other combinations are due to vital actions which are not yet understood. Thus some bioplasm produces muscle; other particles originate nerve structure, cartilage, bone, connective tissue, etc. Many chemical changes occur also in formed material after its production. It may become dry or fluid, or split up into gaseous or soluble substances as soon as produced. Imperfect oxidation may lead to the formation of fatty matters, uric acid, oxalates, sugar, etc. At the earliest period of development, the formed material consists principally of albuminous and fatty matters, with chlorides, alkaline and earthy phosphates. At a later period gelatin, with amyloid or starchy matter, is produced.*

6. *Varieties in the Form and Function of Bioplasm.*—Mutability of shape is characteristic of amœboid cells, and no conclusions can be drawn from their appearance after death. Where numbers of them are accumulated, they are flattened by mutual pressure so as to appear polyhedral, laminated, or prismatic. The upper layers of laminated epithelium are usually flattened. Where cells line the interior of cavities in a single layer, they form

---

* See Physiological Anatomy, by Todd, Bowman, and Beale.

plates of different shape (endothelial cells), or cells in which the long axis predominates (cylindrical epithelium), or forms which are intermediate between plates and cylinders. Some cells appear ramified or stellate, as in the cells from the pith of a rush, bone-cells, and corpuscles of the cornea. Others may become extraordinarily elongated, as in the formation of fibre, muscle, etc. Some cells are provided with cilia, which are limited to one portion of the surface, and project their free extremities into the cavity which they line. Dr. Beale considers the cilia to be formed material, and their movements not vital, but a result of changes consequent on vital phenomena.

Every living organism, plant, animal, or man, begins its existence as a minute particle of bioplasm. Every organic form, leaves, flowers, shells, and all varieties of animals; and every tissue, cellular, vesicular, hair, bone, skin, muscle, and nerve, originates by subdivision and multiplication and change of bioplasm, and the transformation or metamorphosis of bioplasm into formed material. It is evident, therefore, that there are different kinds of bioplasm indistinguishable by physics and chemistry, but endowed with different powers.*

7. *Cell-Genesis.*—Schleiden first showed that the embryo of a flowering plant originates in a nucleated cell, and that from such cells all vegetable tissues are developed. The original cells were formed in a plasma or blastema, commonly found in pre-existing cells, the nuclei first appearing and then the cell-membrane. These views were applied by Schwann to animal structure. The latter believed that the extra-cellular formation of cells, or their origin in a free blastema, was most frequent in animals. The researches of succeeding physiologists have, however, led to a general belief that all cells originate from other cells.

---

* Beale's Bioplasm.

The doctrine of spontaneous generation or *abiogenesis* has been the object of considerable research, but the brilliant experiments of Pasteur have shown that when all access of living organisms into fluids is prevented, no development of such organisms can be proved in any case to occur. If the access of air, for instance, to a liquid which has been boiled, is filtered through a plug of cotton-wool, no living forms will appear in the liquid, but on examination, such forms will be found in considerable numbers in the cotton-wool, proving the presence of these forms or their germs in the external air. Recent experiments also render it probable that some cell-germs are indestructible by a heat far exceeding that of boiling water.

There are three forms of cell-multiplication, by fission, by germination or budding, and by internal division. The latter mode is termed endogenous. In it new cells are produced within a parent-cell by the separation of the bioplasm into a number of distinct masses, each of which may become a new cell, as in the fecundated ovum.

Fission, or the division by cleavage of a parent-cell into two or four parts, may be regarded as a modification of endogenous cell-multiplication. A good example of it may be seen in cartilage.

Budding or germination consists in the projection of a little process or bud from the mass of bioplasm, which is separated by the constriction of its base, and becomes an independent cell.

8. *Reproduction in the higher organisms* consists essentially of the production of two distinct elements, a germ-cell or ovum, and a sperm-cell or spermatozoid, by the contact of which the ovum is enabled to develop a new individual. Sometimes these elements are produced by different parts of the same organism, in which case the sexes are said to be united, and the individual is called hermaphrodite, androgynous, or monœcious. In other in-

stances the sexes are distinct, and the species are called diœcious.

9. *The alternation of generations* is a term given in biology to express a form of multiplication which occurs in some of the more simple forms of life. It consists really of the alternation of a true sexual generation with the phenomenon of budding. Thus a fern spore gives rise, by budding and cell-division, to a prothallium; this produces archegonia and antheridia, as the sexual elements are called, and the embryo which results from sexual union produces not a prothallium but a fern. This phenomenon is better seen in the *Hydrozoa*. In these the egg produces a minute, ciliated, free-swimming body, which attaches itself, becomes tapering, develops a mouth and tentacles, and is known as the *Hydra tuba*. This multiplies itself, and produces extensive colonies by germination, but under certain circumstances divides by fission and produces *Medusæ*, which develop ova.

10. *Parthenogenesis* designates the production of new individuals by virgin females without the intervention of a male. It has also been applied to germination and fission in sexless beings. In the *Aphides*, ova are hatched in spring, but ten or more generations are produced viviparously and without sexual union throughout the summer. In autumn, however, the final brood are winged males and wingless females, from whose union ova are produced in the ordinary manner.

11. *Transformation and metamorphosis* relate to certain changes or variations of development in the structure and life history of an individual. Thus an insect is an egg or *ovum*, a caterpillar or *larva*, a pupa or *chrysalis*, and an *imago* or perfect insect, and these changes of condition and structure constitute its development. Much difficulty is caused by the phenomena of metamorphosis in assigning the place of different species, transformations being often mistaken for specific differences. It was formerly

supposed that every animal passed through, in its development, a series of stages in which it resembled the inferior members of the animal scale, and systems of zoology were proposed to be founded on this dream of embryology. Careful research, however, has shown that larval changes present many variations. In some the young exhibit the conditions of adults of lower animals. Thus the *Eolis*, a univalve shell fish, in its young state has all the characteristics of a *Pteropod*, a free-swimming mollusk. Sometimes development is retrograde, and the adult is a degraded form as compared with the larva, thus setting at nought all our theories, and teaching us that it is better to observe than to imagine.

12. *Discrimination of Living Forms.*—We have seen, section 6, that there are different kinds of living matter endowed with different powers. We have also seen, section 7, how varied are the forms of multiplication. Yet when we come to discriminate between animal and vegetable life, we find it exceedingly difficult, especially in their more simple forms. Neither form, nor chemical composition, nor structure, nor motive power, affords sufficient grounds for discrimination. Yet when we consider the functions of bioplasm in its varied forms, we may conveniently group all living beings in three great divisions, viz., *fungi*, *plants*, and *animals*.

The bioplasm of the plant finds its pabulum in merely inorganic compounds, while that of the animal is prepared for it, directly or indirectly, by the vegetable. The function of fungi appears to be the decomposition of the formed material of plants and animals by the means of fermentation or putrefaction, since these latter processes are dependent on the presence of fungi. Thus by bioplasm are the structures of plants and animals reared from inorganic materials, and by bioplasm are they broken down and restored to the inanimate world.

CHAPTER X.

THE MICROSCOPE IN VEGETABLE HISTOLOGY AND BOTANY.

HISTOLOGY (from ἱστος, a tissue) treats of formed material, or the microscopic structure resulting from the transformation of germinal or living matter. The nature of this transformation is partly physical and partly vital, and, as already stated, is often so complex as to baffle all chemical analysis. Some light, however, has been thrown on this subject by the modification of ordinary crystalline forms when inorganic particles aggregate in the presence of certain kinds of organic matter. To this mode of formation the name of *molecular coalescence* has been given. Mr. Rainey and Professor Harting contemporaneously experimented with solutions of organic colloids, and found that the crystallization of certain lime salts, as the carbonate, was so modified by such solutions as to resemble many of the calcareous deposits found in nature. These researches leave little doubt but that a majority of calcareous and silicious organic forms may be thus accounted for. Such changes are rather physical than vital.

Cell-substance in Vegetables.—The protoplasm or bioplasm in vegetable-cells cannot be distinguished from animal "sarcode" or protoplasm except by the nature of the pabulum or aliment necessary to its nutrition. The vegetable, under the stimulus of light, decomposes carbonic acid, and acquires a red or green color from the compounds which it forms, while the animal requires nutriment from pre-existing organisms. Yet this definition fails to apply to fungi, which resemble primitive animals even in this respect. So difficult is it to discriminate that the simpler forms of vegetables have often been classed by naturalists among animals, and *vice versâ*. Amœboid movements have been observed in the bioplasm of vegetable-cells, especially in the *Volvox*, and some have considered it probable that an organism may live a truly vegetable life at one period and a truly animal life at another.

Analogous to amœboid movements, is the motion of bioplasmic fluid in the interior of undoubtedly vegetable cells. This movement is called *cyclosis*, and may be detected under the microscope by the granules or particles which the current carries with it in the transparent cells of *Chara*, *Vallisneria*, etc., and in the epidermic hairs of many plants, as *Tradescantia*, *Plantago*, etc. (Plate VI, Fig. 91).

The bioplasm of plants may be stained with carmine solution without affecting the cell-wall or other formed material.

Cell-wall or Membrane.—Plants, whether simple or complicated in structure, are but cells or aggregations of cells. In the simplest vegetables or *Protophytes*, each cell lives as it were an independent life, performing every function; while in the higher plants, as the palm or oak, the cells undergo special modifications, and serve various functions subsidiary to the life of the plant as a whole.

Cell-membrane, or the envelope of formed material, was formerly thought to be composed of two layers, to the inner one of which the name of *primordial utricle* was given, but this is now considered to be but the external surface of the bioplasm or germinal matter.

The chemical nature of cell-membrane is nearly identical with starch, being composed of *cellulose*. The presence of cellulose may be shown by the blue color which is produced by applying iodine and sulphuric acid, or the iodized solution of chloride of zinc.

Endosmose will take place in cell-membrane, allowing solutions to pass through, as pabulum, and the manner of this passage may in some instances determine the subsequent deposit of formed material. Sometimes actual pores are left in the membrane, as in *Sphagnum* (Plate VI, Fig.

92). The walls of vegetable-cells are often thickened by deposit. If this is in isolated patches, the cells are called *dotted* (Plate VI, Fig. 93), and it is sometimes difficult to distinguish them from porous cells. Many cells have a spiral fibre (Plate VI, Fig. 94), which appears to have been detached from the outer membrane. In the seeds of *Collomia*, etc., the cell-wall is less consolidated than the deposit, so that on softening the cells by water, the spiral fibres suddenly spring out, making a beautiful object for a half-inch object-glass (Plate VI, Fig. 95).

The tendency of formed material to arrange itself in a spiral is seen in the endochrome of many of the simpler plants, as *Zygnema*, and the cell-wall sometimes tears most readily in a spiral direction.

If the spiral deposit is broken and coalesces at some of its turns, it forms an annulus or ring. Some cells show both rings and spirals.

For the production of a spiral movement or growth, another force is needed in addition to the centripetal and centrifugal forces which are necessary for curvilinear motion. The centripetal point must be carried forward in space by a progressive force. When we consider that a spiral form is so frequently seen in morphology, that the secondary planets move in spirals round their primaries, and that even in distant nebulæ the same law prevails, we are struck with the unity of plan which is exhibited throughout the universe, and can scarcely fail to observe that even a microscopic cell shows the tracings of the same divine handiwork which swings the stars in their courses.

Sclerogen — Ligneous Tissue. — Sometimes the deposit within the cell-wall is of considerable thickness, and often in concentric rings, through which a series of passages is left so that the outer membrane is the only obstacle to the access of pabulum, as in the stones of fruit, gritty tissue of the pear, etc. (Plate VII, Fig. 96). The nature of this deposit is similar to cellulose, although often contain-

Fig. 91.

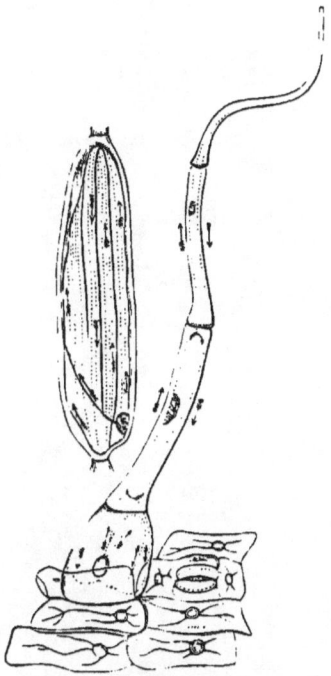

Circulation of fluid in hairs of *Tradescantia Virginica.*

Fig. 92.

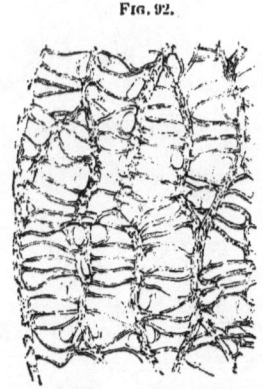

Portion of the leaf of *Sphagnum.*

Fig. 93.

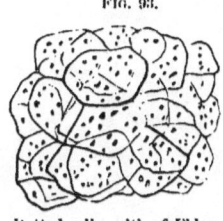

Dotted cells—pith of Elder.

Fig. 94.

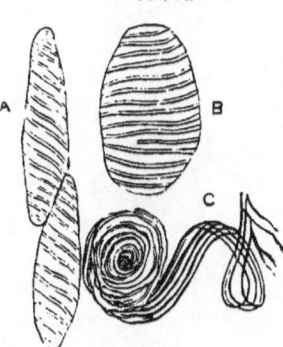

Spiral cells :—A, Balsam ; B, C, Pleurothallis.

Fig. 95.

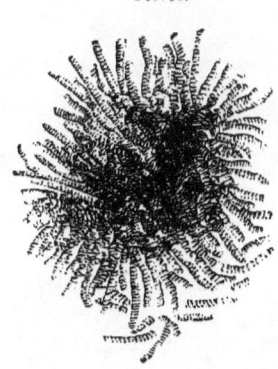

Spiral fibres of seed-coat of *Collomia.*

ing resinous and other matters. Woody fibre or ligneous tissue is quite similar, save that the cells have become elongated or fusiform, and when completely filled up with internal deposit, fulfil no other purpose than that of mechanical support (Plate VII, Fig. 97). The woody fibres of the *Coniferæ* exhibit peculiar markings, which have been called *glandular* (Plate VII, Fig. 98). In these the inner circle represents a deficiency of deposit as in other porous cells, while the outer circle is the boundary of a lenticular cavity between the adjacent cells. This arrangement is so characteristic as to enable us to determine the tribe to which a minute fragment, even of fossil wood, belonged.

Spiral Vessels.—If spiral cells are elongated, or coalesce at their ends, they become vessels, some of which convey air and some fluid (Plate VII, Fig. 99). As in cells, the want of continuity in the spiral fibre sometimes produces rings, when the duct is called *annular*. In other instances the spires are still more broken up by the process of growth, so as to form an irregular network in the duct, which is then said to be *reticulated*. A still greater variation in the deposit produces *dotted ducts*. Not infrequently we find all forms in the same bundle of vessels.

Laticiferous Vessels (Plate VII, Fig. 100).—These convey the milky juice or *latex* of such plants as possess it, as the Euphorbiaceæ, india-rubber plant, etc., and differ from the ducts above described by their branching, so as to form a network, while ducts are straight and parallel with each other.

The laticiferous vessels resemble the capillary vessels of animals, while the spiral ducts remind us of the trachea of insects.

Siliceous Structures.—The structures of many plants, especially the epidermis, often become so permeated with a deposit of silica, that a complete skeleton is left after the soft vegetable matter is destroyed. The frustules of

Diatoms have in this way been preserved in vast numbers in the rocky strata of the earth. The markings on these siliceous shells are so delicate as to be employed as a test of microscopic power and definition. In a species of *Equisetum* or Dutch rush, silica exists in such abundance that the stems are sometimes employed by artisans as a substitute for sand-paper. If such a stem is boiled and macerated in nitric acid until all the softer parts are destroyed, a cast of pure silica will exhibit not only the forms of the epidermic cells, but details of the stomata or pores. The same also is true of the husk of a grain of wheat, etc., in which even the fibres of the spiral vessels are silicified. The stellate hairs of the siliceous cuticle from the leaf of *Deutzia scabra* forms a beautiful polariscope object.

Formed Material within Vegetable Cells.

1. *Raphides.*—These are crystalline mineral substances, principally oxalate, citrate, and phosphate of lime. They occur in all parts of the plant, sometimes in the form of bundles of delicate needles, sometimes in larger crystals, and sometimes in stellate or conglomerate form. Mr. E. Quekett produced such forms artificially by filling the cells of rice-paper with lime-water under an air-pump, and then placing the paper in weak solutions of phosphoric or oxalic acid.

2. *Starch.*—This performs in plants a similar function to that of fat in animals, and is a most important ingredient in human food, since two-thirds of mankind subsist almost exclusively upon it. It is found in the cells of plants in the form of granules or secondary cells. Each granule under the microscope shows at one extremity a circular spot or *hilum*, around which are a number of curved lines, supposed to be wrinkles in the cell-membrane. When starch is boiled in water, this membrane bursts and

Fig. 96.

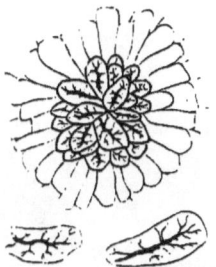

Gritty tissue—Pear.

Fig. 99.

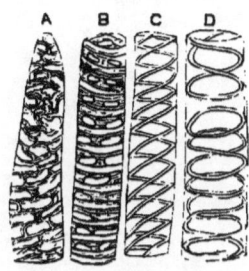

Spiral vessels:—A, reticulated; B, old vessel, with perforations; C, D, spiral vessels, becoming annular.

Fig. 97.

Wood-fibre—flax.

Fig. 100.

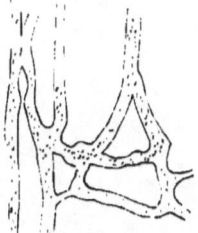

Lactiferous vessels.

Fig. 98.

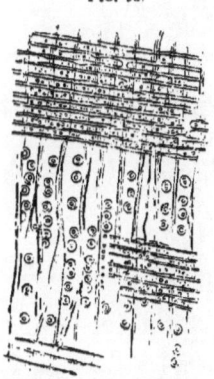

Section of *Coniferous Wood* in the direction of the fibres.

Fig. 101.

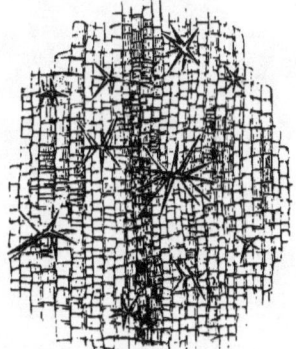

Cubical parenchyma, with stellate cells, from petiole of *Nuphar lutea*.

the amylaceous matter is dissolved. Iodine stains starch blue. Starch shows in the polariscope a black cross in each grain, changing to white as the prism is revolved.

3. *Chlorophyll* is the green coloring matter of plants. It is usually seen in the form of granules of bioplasm in the interior of cells. These green granules yield their chlorophyll to alcohol and ether. It seems to be necessary to nutrition, since green plants under the stimulus of light break up carbonic acid into oxygen and carbon, the latter of which is absorbed.

The red and yellow color of autumn leaves is owing to the chemical metamorphosis of chlorophyll, as also is the red color of many of the lower Algæ, etc. In the latter it seems to be in some way connected with the vital processes.

4. *The coloring matter of flowers* is various, and ordinarily depends on the colored fluid contained in cells subjacent to the epidermis, although sometimes it is in the form of solid corpuscles. White patches on leaves, etc., arise from absence of chlorophyll.

5. *Milky juices* are true secretions contained in the laticiferous ducts. The juice of the dandelion, caoutchouc or india-rubber, which is the concrete juice of the *Ficus elastica*, and gutta-percha, from *Isonandra gutta*, are examples.

6. *Fixed oils* are found in the cells of active tissues, and notably in seeds, where they serve to nourish the embryo. Cocoanut, palm, castor, poppy, and linseed oils are examples.

7. *Volatile oil*, sometimes called essential oil, is chiefly found in glandular cells and hairs of the epidermis. Many of them yield a resinous substance by evaporation.

8. *Camphor* is analogous to volatile oil, although solid at ordinary temperatures. It abounds in the Lauraceæ.

9. *Resin*, *wax*, and *tallow* are also found in plants. The bloom of the plum and grape is due to wax.

10. *Gum* is a viscid secretion. What is called gum tragacanth, is said to be partially decomposed cell-membrane, and is allied to amyloid matter.

Forms of Vegetable Cells.—From the account given in the chapter on biology, page 123, it is evident that the form of cells is quite varied, and often depends on the amount of pressure from aggregation, yet function also has much to do in the determination of shape. Thus while most elongated cells are lengthened in the direction of plant-growth, in which is least resistance, the medullary rays of Exogenous stems are elongated in a horizontal direction. Some cells are cubical, as in the leaves of the yellow water-lily, *Nuphar lutea* (Plate VII, Fig. 101). Others are stellate, as in the rush (Plate VIII, Fig. 102). In many tissues are large cavities or air-chambers altogether void of cells, and in leaves such cavities communicate with the external air by means of *stomata* or pores (Plate VIII, Fig. 103), which are usually provided with peculiar cells for contracting or widening the orifice.

The Botanical Arrangement of Plants.—Considered with reference to their general structure, plants are divided by botanists into *cellular* and *vascular*. The first of these classes is of greatest interest to the microscopist, as embracing the minuter forms of vegetable life.

The classification and natural grouping of plants is yet far from being perfect, although microscopic examinations have largely contributed to an orderly arrangement of the multitudinous varieties in this field of research. In the present work we propose only a brief outline of typical subjects of interest, with the methods of microscopic examination.

Fungi.—At page 127 it was stated that all living beings may be grouped in three divisions, fungi, plants, and animals. Botanists generally class fungi among cellular flowerless plants. They cannot assimilate inorganic food as other plants, but live upon the substance of animal or

Fig. 102.

Section of cellular parenchyma of *Rush*.

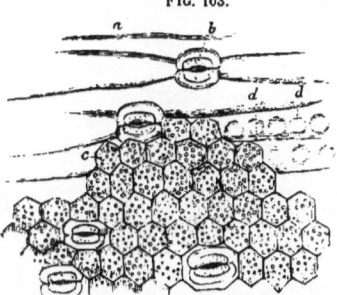

Portion of the cuticle of the leaf of the *Iris Germanica*, torn from its surface.

Fig. 104.

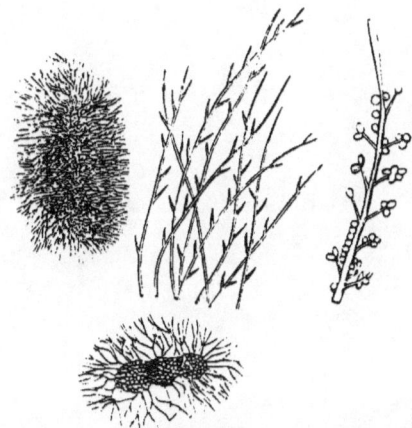

Botrytis bassiana.

Fig. 105.

Cells from the petal of the Geranium (*Pelargonium*).

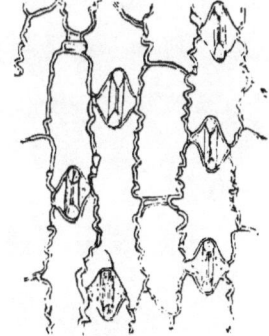

Cuticle of leaf of Indian Corn (*Zea mais*).

vegetable tissue. They also differ from ordinary vegetables by the total absence of chlorophyll or its red modification. A large number of this strange class are microscopic, and require high powers for their observation. Recent investigations show that individual fungi are developed in very dissimilar modes, and are subject to a great variety of form, rendering it probable that those which seem most simple are but imperfectly developed forms. Amœboid motions also in the cell-substance of certain kinds of fungi, and the projection of threads of bioplasm, show a great resemblance to some of the lower forms of animal life, as the *Rhizopods*.

All fungi exhibit two well-defined structures, a *mycelium* or vegetative structure, which is a mass of delicate filaments or elongated cells; and a *fruit* or reproductive structure, which varies in different tribes. In *Torula*, one or more globular cells are produced at the ends of filaments composed of elongated cells; these globules drop off and become new mycelia. The "yeast plant," or *Torula cerevisia* (Plate IX, Fig. 107), receives its name from its habitat. Fermentation depends upon its presence, as putrefaction does upon the minute analogous bodies called *Bacteria* and *Vibriones*. Bacteria are minute, moving, rod-like bodies, sometimes jointed; and vibriones are moniliform filaments, having a vibratile or wriggling motion across the field of view in the microscope. The researches of Madame Luders render it probable that the germs of fungi develop themselves into these bodies when sown in water containing animal matter, and into yeast in a saccharine solution. The universal diffusion of sporules of fungi in the atmosphere readily accounts for their appearance in such fluids, and Pasteur's experiments are quite conclusive.

The minute molecules called *microzymes*, present in various products of disease, as the vaccine vesicle, fluid of glanders, etc.; the minute corpuscles which cause the disease among silkworms called "pebrine;" etc.; have a strong analogy in their rapid multiplication to the yeast-cells.

The sporules of any of the ordinary moulds, as *Penicillium*, *Mucor*, or *Aspergillus*, will develop into yeast-cells in a moderately warm solution of cane-sugar, showing how differently the same type of bioplasm may develop under different conditions. The term *polymorphism* has been given to this phenomenon. Very many species, and even genera, so called, may after all be only varieties of the same kind of organism.

In many morbid conditions of the skin and mucous membranes, there is not only an alteration or morbid growth of the part, but a vegetation of fungi. Thrown-off scales of epithelium from the mouth and fauces exhibit fibres of *leptothrix*, and the false membrane of diphtheria, as well as the white patches of aphtha or thrush, show the mycelia and spores of fungi. The disease in silkworms called muscardine is due to a fungus, the *Botrytis bassiana* (Plate VIII, Fig. 104), whose spores enter and develop in the air-tubes. The filamentous tufts seen about dead flies on window-panes, etc., arise from a similar growth of *Achyla*. In certain Chinese or Australian caterpillars, this sort of growth becomes so dense as to give them the appearance of dried twigs. Even shells and other hard tissues may become penetrated by fungi. The dry rot in timber is a form of fungus.

The mildew which attacks the straw of wheat, etc., arises from the *Puccinia graminis*, whose spores find their way through the stomata or breathing pores of the epidermis. *Rust*, and *smut*, and *bunt*, originate in varieties of *Uredo*. The "vine disease" and the "potato disease," as they are called, have similar origin.

Various methods have been proposed to destroy fungi in growing plants, but it must be remembered that the function of these organisms is chiefly to remove formed

PLATE IX.

Fig. 107

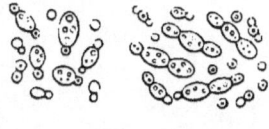

Torula Cerevisiæ, or Yeast-Plant.

Fig. 109.

Germ and Sperm-cells in Achyla.

Fig. 108.

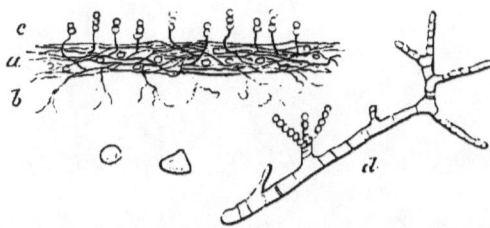

Development of fungi: A, mycelium; B, hypha; C, conidiophores; D, a magnified branch.

Fig. 110.

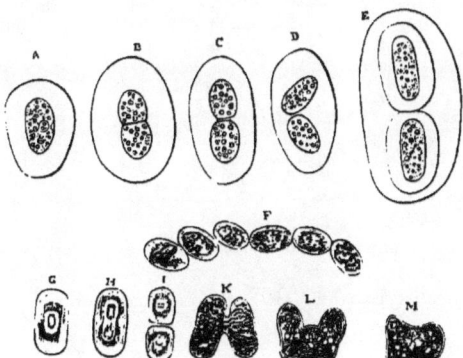

Various phases of development of *Palmoglæa macrococca*.

material in a state of decay, which is more or less complete. The prevalence of atmospheric changes, variations in light, heat, moisture, and electricity, etc., have much to do in predisposing vegetable as well as animal tissues to disease and producing epidemics. The agriculturist, therefore, as well as the physician, must discriminate between those diseased conditions which provide a habitat for fungi, and the effects produced by the fungi themselves.

Impregnating wood with corrosive sublimate or chloride of zinc has been used to prevent dry rot in wood, and soaking seeds in alkaline solutions or sulphate of copper is said to remove smut and similar fungus spores.

The development of fungi is from spores or *conidia*. Plate IX, Fig. 107, represents the Torula vegetating by the budding of its spores. These buds rapidly fall off and become independent cells. In other varieties self-division gives rise to the *mycelium*, a mass of fibres often interlaced so as to form a sort of felt. Some branches of this mycelium (*hyphæ*) hang down, while others rise above the surface (*conidiophores*) and bear conidia, which fall off and develop into new hyphæ (Plate IX, Fig. 108). In the "blight" of the potato the mycelium is loose, and the hyphæ ramify in the intercellular spaces and give off projections into the cells of the plant. The conidia germinate by bursting the sac which contains them, putting forth cilia, moving awhile, then resting and enveloping themselves with membrane and growing into hyphæ. In the autumn, parts of the hyphæ assume special functions. One part develops a spherical mass called *oogonium*, while another becomes a smaller mass or *antheridium*. When the first is ripe, it is penetrated by the latter, and the bioplasms of each are fused together. The antheridium then decays, while the oogonium grows and becomes an *oospore*, in which the bioplasm divides and subdivides. Next season each segment escapes ciliated, and moves about till it finds a place to germinate. In Achyla two sacs are formed, one of which contains "germ-cells," and the other *antherozoids* or "sperm-cells." When both are ripe the sac opens, and the ciliated antherozoids pass into the neighboring sac and fertilize its contents (Plate IX, Fig. 109).

In other fungi the reproductive cells are undistinguishable from the rest, and the coalescence takes place in a new cell formed by the union of the other two.

Mr. Berkeley divides fungi into six orders, as follows:

1. *Hymenomycetes* or *Agaricoideæ* (*Mushrooms*, etc.).—Mycelium flocuose, inconspicuous, bearing fleshy fruits which expand so as to expose the hymenium or sporiferous membrane to the air. Spores generally in fours on short pedicles.

2. *Gasteromycetes* or *Lycoperdoideæ* (*Puff balls*, etc.).—Fruit globular or oval, with convolutions covered by the hymenium, which bears the spores in fours on distinct pedicles. The convolutions break up into a pulverulent or gelatinous mass.

3. *Coniomycetes* or *Uredoideæ* (*Smuts*, etc.).—Mycelium filamentous, parasitic. Microscopic fructification of sessile or stalked spores in groups, sometimes septate.

4. *Hyphomycetes* or *Botrytoideæ* (*Mildews*, etc.).—Microscopic. Mycelium filamentous, epiphytic, with erect filaments bearing terminal, free, single, simple, or septate spores.

5. *Ascomycetes* or *Helvelloideæ* (*Truffles*, etc.).—Mycelium inconspicuous. Fruit fleshy, leathery, horny, or gelatinous, lobed, or warty, with groups of elongated sacs (*asci* or *thecæ*) in which the spores (generally eight) are developed.

6. *Physomycetes* or *Mucoroideæ* (*Moulds*).—Mycelium (microscopic) filamentous, bearing stalked sacs containing numerous minute sporules.

Protophytes, or primitive plants, afford many forms and groups of great interest to the microscopist as well as to the biologist. The plan of the present work permits us only to indicate a few particulars, the details of which would form a volume of considerable size.

The Algæ are divided into three orders: I. *Rhodospermeæ* or *Floridæ* (*Red-spored Algæ*). Marine plants, with a leaf-like or filamentous rose-red or purple thallus. II. *Melanosporeæ* or *Fucoideæ* (*Dark-spored Algæ*). Marine. Thallus leaf-like, shrubby, cord-like, or filamentous, of olive-green or brown color. III. *Chlorosporeæ* or *Confervoideæ* (*Green-spored Algæ*). Plants marine or fresh water, or growing on damp surfaces. Thallus filamentous, rarely leaf-like, pulverulent, or gelatinous. These have been subdivided into families, viz.:

I. *Rhodospermeæ.*—1. Rhodomelaceæ. 2. Laurenciaceæ. 3. Corallinaceæ. 4. Delesseriaceæ. 5. Rhodymeniaceæ. 6. Cryptonemiaceæ. 7. Ceramiaceæ. 8. Porphyraceæ.

II. *Melanosporeæ.*—1. Fucaceæ. 2. Dictyotaceæ. 3. Cutleriaceæ. 4. Laminariaceæ. 5. Dictyosiphonaceæ. 6. Punctariaceæ. 7. Sporochnaceæ. 8. Chordariaceæ. 9. Myrionemaceæ. 10. Ectocarpaceæ.

III. *Chlorosporeæ.*—1. Lemaneeæ. 2. Batrachospermeæ. 3. Chœtophoraceæ. 4. Confervaceæ. 5. Zygnemaceæ. 6. Œdogoniaceæ. 7. Siphonaceæ. 8. Oscillatoriaceæ. 9. Nostochaceæ. 10. Ulvaceæ. 11. Palmellaceæ. 12. Desmidiaceæ. 13. Diatomaceæ. 14. Volvocineæ.

For fuller information, we refer to the *Micrographic Dictionary* by Griffith and Henfrey.

In the family of *Palmellaceæ* we find the simplest forms of vegetation in the form of a powdery layer of cells, or a slimy film, or a membranous frond. The green mould on damp walls and the red snow of alpine regions are examples.

In the green slime on damp stones, etc., is found the *Palmoglœa macrococca*. The microscope shows it to consist of cells containing chlorophyll, surrounded by a gelatinous envelope. These cells multiply by self-division.

Sometimes a conjugation or fusion of cells occurs, and the product is a *spore* or primordial cell of a new generation (Plate IX, Fig. 110). During conjugation oil is produced in the cells, and the chlorophyll disappears or becomes brown, and when the spore vegetates, the oil disappears and green granular matter takes its place. This is analogous to the transformation of starch into oil in the seeds of the higher plants.

Most of the lower forms of vegetable life pass through what is called the motile condition, which depends on the extension of the bioplasm into thread-like filaments, whose contractions serve to move the cell through the water. Many of these forms were formerly mistaken for animalcules, and the transformation of a portion of green chlorophyll into the red form was represented as an eye. The multiplication of the "still" cells is by self-division, as in *Palmoglœa*, but after this has been repeated about four times, the new cells become furnished with cilia and pass into the "motile" condition, and their multiplication goes on in different ways, as by binary or quaternary segmentation, or the formation of a compound, mulberry-like mass, the ciliated individual cells of which, becoming free, rank as *zoospores* (Plate X, Fig. 111).

The *Volvox* is a beautiful example of the composite motile form of elementary vegetation. It is found in fresh water, and consists of a hollow pellucid sphere, studded with green spots, connected together often by green threads. Each of these spots has two cilia, whose motions produce a rolling movement of the entire mass. Within the sphere there are usually from two to twenty smaller globes, which are set free by the bursting of the original envelope. Sometimes one of the masses of endochrome enlarges, but instead of undergoing subdivision becomes a moving mass of bioplasm, which cannot be distinguished from a true *Amœba* or primitive animal cell.

The *Desmidiaceæ* are a family of minute green plants

PLATE X.

Fig. 111.

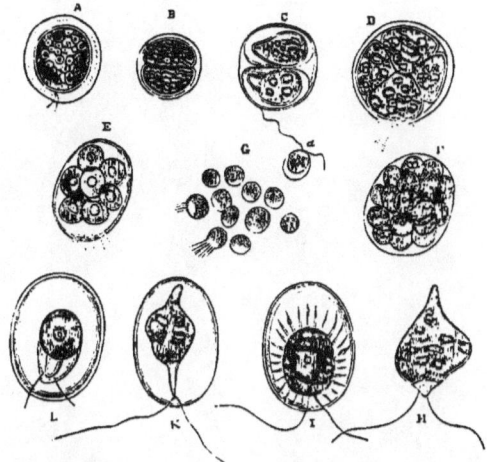

Various phases of development of *Protococcus pluvialis*.

Fig. 112.

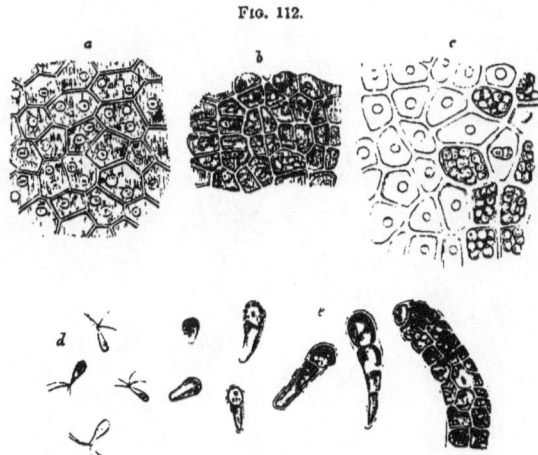

Formation of Zoospores in *Phycoseris gigantea* (Ulva latissima).

of great interest. Generally the cells are independent, but a filament is sometimes formed by binary subdivision. Their symmetrical shape, and frequently spinous projections and peculiar movements, render them beautiful objects. By conjugation a spore-cell or *sporangium* is produced, which in some species is spinous, and resembles certain fossil remains in flint, which have been described as animalcules under the name of *Xanthidia*.

The family of *Diatomaceæ* affords more occupation to microscopists than other protophytes. Like the *Desmids*, they are simple cells with a firm external coating, but in Diatoms this coating is so penetrated with silex, that a cast of the frustule is left after the removal of the organic matter. Reference has already been made to the number of these organisms in a fossil state, as well as to their utility as tests of the defining power of microscopic object-glasses.

Some species inhabit the sea, and others fresh water. They are so numerous that scarcely a ditch or cistern is free from specimens, and they multiply so rapidly as to actually diminish the depth of channels and block up harbors. They may be sought for in the slimy masses attached to rocks and plants in water, in the scum of the surface, in mud or sand, in guano, in the stomachs of molluscs, etc., and on sea-weeds.

To separate the shields or siliceous frustules from foreign matter, either fresh or fossil, they should be washed several times in water, and the sediment allowed to subside. The deposit should then be treated in a test-tube with hydrochloric acid, sometimes aided by heat. This should be repeated as often as any effect is produced, and then the sediment should be boiled in strong nitric acid, and washed several times in water. They may be mounted dry or in balsam.

The classification of Diatoms is not yet perfected, but Muller's type slides, containing from one hundred to five hundred characteristic forms, is a valuable assistance. The following table, from the *Micrographic Dictionary,* gives an analysis of tribes and genera: Fr. denotes the frustules in front view; v. the valves; granular striæ means striæ resolvable into dots; and continuous striæ signify costæ or canaliculæ.

A. *Frustules not contained in a Gelatinous Mass or Tube.*

TRIBE I. STRIATÆ.—Frustules usually transversely striate, but neither vittate nor areolate.

† *Valves without a Median Nodule.*

COHORT 1. EUNOTIEÆ.—Fr. arcuate, single, or united into a straight filament.

1. *Epithemia.*—Fr. single or binate, with transverse or slightly radiant striæ, some continuous; no terminal nodules; aquatic and marine.

2. *Eunotia.*—Fr. single or binate; v. with slightly radiant granular striæ and terminal nodules; aquatic.

3. *Himantidium.*—Fr. as in *Eunotia,* but united into a filament; striæ parallel, transverse; aquatic.

COHORT 2. MERIDEÆ.—Fr. cuneate, single, or united into a curved or spinal band; v. with continuous or granular striæ.

4. *Meridion.*—Fr. cuneate, united into a spiral band; striæ continuous; aquatic.

5. *Eucampia.*—Fr. united into an arched band; v. punctate; marine.

6. *Oncosphenia.*—Fr. single, cuneate, uncinate at the narrow end; striæ granular; aquatic.

COHORT 3. FRAGILLARIEÆ.—Fr. quadrilateral, single, or united into a filament or chain; v. with continuous or granular striæ.

7. *Diatoma.*—Fr. linear or rectangular, united by the

angles so as to form a zigzag chain; striæ continuous; aquatic and marine.

8. *Asterionella.*—Fr. adherent by adjacent angles into a star-like filament; v. inflated at one or both ends; aquatic.

9. *Fragillaria.*—Fr. linear, united into a straight, close filament; striæ granular, faint; aquatic and marine.

10. *Denticula.*—Fr. linear, simple, or binate, rarely more united; striæ continuous; aquatic.

11. *Odontidium.*—As *Denticula*, but fr. forming a close filament; aquatic and marine.

COHORT 4. MELOSIREÆ.—Fr. cylindrical, disk-shaped or globose; v. punctate, or often with radiate continuous or granular striæ.

12. *Cyclotella.*—Fr. disk-shaped, mostly solitary; v. with radiate marginal striæ; aquatic.

13. *Melosira.*—Fr. cylindrical or spherical, united into a filament; v. punctate, or with marginal radiate granular striæ; aquatic and marine.

14. *Podosira.*—Fr. united in small numbers, cylindrical or spherical, fixed by a terminal stalk; v. hemispherical, punctate; marine.

15. *Mastogonia.*—Fr. single; v. unequal, angular, mammiform, circular at base, without umbilical processes; angles radiating; fossil.

16. *Podoliscus.*—Fr. single or united, with a marginal stalk; v. circular, convex.

17. *Pyxidicula.*—Fr. single or binate, free or sessile; v. convex; aquatic and marine.

18. *Stephanodiscus.*—Fr. single, disk-shaped; v. circular, equal, punctate, or striate, with a fringe of minute marginal teeth; aquatic.

19. *Stephanogonia.*—Fr. as in *Mastogonia*, but ends of valves truncate, angular, and spinous; fossil.

20. *Hercotheca.*—Fr. single, turgid laterally; v. with marginal free setæ.

21. *Goniothecium.*—Fr. single, constricted in the middle, suddenly attenuate and truncate at the ends (hence appearing angular).

COHORT 5. SURIRELLEÆ.—Fr. single or binate, quadrilateral, oval, or saddle-shaped, sometimes constricted in the middle; v. with transverse or radiating continuous or granular striæ, interrupted in the middle, or with one or more longitudinal rows of puncta; often keeled.

22. *Bacillaria.*—Fr. prismatic, straight, at first forming a filament; v. with a median longitudinal row of puncta; marine.

23. *Campylodiscus.*—Fr. single, free, disk-shaped; v. curved or twisted (saddle-shaped); aquatic and marine.

24. *Doryphora.*—Fr. single, stalked; v. lanceolate or elliptical, with transverse granular striæ.

25. *Podocystis.*—Fr. attached, sessile; v. with a median line, transverse continuous, and intermediate granular striæ.

26. *Nitzschia.*—Fr. free, single, compressed, usually elongate, straight, curved, or sigmoid, with a not-median keel, and one or more longitudinal rows of puncta; aquatic and marine.

27. *Sphinctocystis (Cymatopleura).*—Fr. free, single, linear, with undulate margins; v. oblong or elliptical, sometimes constricted in the middle; aquatic.

28. *Surirella.*—Fr. free, single, ovate, elliptical, oblong, cuneate, or broadly linear; v. with a longitudinal median line or clear space, margins winged, and with transverse or slightly radiating continuous striæ; aquatic and marine.

29. *Synedra.*—Fr. prismatic, rectangular, or curved; at first attached to a gelatinous-lobed cushion, often becoming free; v. linear or lanceolate, usually with a median pseudo-nodule and longitudinal line; aquatic and marine.

30. *Tryblionella.*—Fr. free, linear, or elliptical; v. plane, with a median line, transverse striæ, and submarginal or obsolete alæ; aquatic and marine.

31. *Raphoneis.*—*Doryphora* without a stalk.

COHORT 6. AMPHIPLEUREÆ.—Fr. free, single, straight, or slightly sigmoid; v. lanceolate, or linear-lanceolate, with a median longitudinal line.

32. *Amphipleura.*—Characters as above.

†† *Valves with a Median Nodule.*

COHORT 7. COCCONEIDÆ.—Fr. straight or bent, attached by the end or side; v. elliptical, equilateral.

33. *Cocconeis.*—Fr. single, compressed, adnate; v. elliptical, one of them with a median line.

COHORT 8. ACHNANTHEÆ.—Fr. compressed, single, or rarely united into a straight filament, curved, attached by a stalk at one angle; uppermost v. with a longitudinal median line, lower v. the same, and a stauros or transverse line; marine.

35. *Achnanthidium.*—Fr. those of *Achnanthes*, but free; aquatic.

36. *Cymbosira.*—Fr. as *Achnanthes*, solitary or binate, stipitate, and attached end to end; marine.

COHORT 9. CYMBELLEÆ.—Fr. straight or curved, free or stalked at the end; v. inequilateral, not sigmoid.

37. *Cymbella.*—Fr. free, solitary; v. navicular, with a subcentral and two terminal nodules, and a submedian longitudinal line; aquatic.

38. *Cocconema.*—Fr. as *Cymbella*, but stalked; aquatic.

COHORT 10. GOMPHONEMEÆ.—Fr. wedge-shaped, straight, free, or stalked; v equilateral.

39. *Gomphonema.*—Fr. single or binate, wedge-shaped, attached by their ends to a stalk; v. with a median line, and a median and terminal nodules; aquatic.

40. *Sphenella.*—Fr. free, solitary, wedge-shaped, involute; aquatic.

41. *Sphenosira.*—Fr. united into a straight filament; v. wedge-shaped, at one end rounded, suddenly contracted and produced; aquatic.

COHORT 11. NAVICULEÆ.—Fr. free, straight; v. equilateral, or sometimes sigmoid.

42. *Navicula.*—Fr. single, free, straight; v. oblong, lanceolate, or elliptical, with a median line, a central and two terminal nodules, and transversely or slightly radiant lines resolvable into dots; aquatic, marine, and fossil.

43. *Gyrosigma (Pleurosigma).*—Fr. as *Navicula*, but v. sigmoid; aquatic and marine.

44. *Pinnularia.*—Fr. as *Navicula*, but transverse lines continuous; aquatic and marine.

45. *Stauroneis.*—Fr. as *Navicula*, but the median line replaced by a stauros; aquatic and marine.

46. *Diadesmis.*—Fr. as *Navicula*, united into a straight filament; aquatic.

47. *Amphiprora.*—Fr. free, solitary, or in pairs, constricted in the middle; v. with a median keel, and a median and terminal nodules, often twisted; marine.

48. *Amphora.*—Fr. plano-convex, elliptical, oval or oblong, solitary, free or adnate, with a marginal line, and a nodule or stauros on the flat side; aquatic and marine.

TRIBE II. VITTATÆ.—Fr. with vittæ.

† *Valves without a Median Nodule.*

COHORT 12. LICMOPHOREÆ.—Fr. cuneate; vittæ arched.

49. *Licmophora.*—Fr. cuneate, rounded at the broad end, radiating from a branched stalk; vittæ curved (by inflection of upper margins of valves); marine.

50. *Podosphenia.*—Fr. as *Licmophora*, but single or in pairs, sessile on a thick but little branched pedicle; marine.

51. *Rhipidophora.*—Fr. as *Licmophora*, single or in pairs, on a branched stipes; marine.

52. *Climacosphenia.*—Fr. cuneate, rounded at broad end, divided into loculi by transverse septæ or vittæ; marine.

COHORT 13. STRIATELLEÆ.—Fr. tabular or filamentous; vittæ straight (not arched).

53. *Striatella.*—Fr. compound, stalked at one angle; vittæ longitudinal and continuous; v. elliptic-lanceolate, not striated; marine.

54. *Rhabdonema.*—Fr. as *Striatella*, but vittæ interrupted; v. with transverse granular striæ; marine.

55. *Tetracyclus.*—Fr. compound, filamentous; vittæ alternate, interrupted; v. inflated at the middle; striæ transverse, continuous; aquatic.

56. *Tabellaria.*—Fr. united into a filament, subsequently breaking up into a zigzag chain; vittæ interrupted, alternate; v. inflated at middle and ends; aquatic.

57. *Pleurodesmium.*—Fr. tabular, united into a filament, and with a transverse median hyaline band; marine.

58. *Hyalosira.*—Fr. tabular, fixed by a stalk at one angle; vittæ alternate, interrupted, bifurcate at the end; marine.

59. *Anaulus.*—Fr. rectangular, single, compressed, with lateral inflections, giving the valves a ladder-like appearance; marine.

60. *Biblarium.*—Fr. as *Tetracyclus*, but single; fossil.

61. *Terpsinoë.*—Fr. tabular, obsoletely stalked, subsequently connected by isthmi; vittæ transverse, short, interrupted, and capitate; aquatic and marine.

62. *Stylobiblium.*—Fr. compound; v. circular, sculptured with continuous striæ; fossil.

†† *With a Median apparent (pseudo) Nodule.*

63. *Grammatophora.*—Fr. at first adnate, afterwards

forming a zigzag chain; vittæ two, longitudinal, interrupted, and more or less figured; marine.

TRIBE III. AREOLATÆ.—Valves circular, with cell-like (areolar) markings, visible by ordinary illumination.

SUB-TRIBE 1. DISCIFORMES.—Valves alike, without appendages or processes.

COHORT 14. COSCINODISCEÆ.—Valves circular.

64. *Actinocyclus.*—Fr. solitary; v. circular, undulate, the raised portions like rays or bands radiating from the centre, which is free from markings; marine and fossil.

65. *Actinoptychus.*—Fr. as *Actinocyclus*, but radiating internal septæ, as well as rays.

66. *Coscinodiscus.*—Fr. single; v. circular, areolar all over; marine and fossil.

67. *Arachnoidiscus.*—Fr. single; v. circular, not undulate, with concentric and radiating lines, and intermediate areola absent from the centre (pseudo-nodule); marine and fossil.

68. *Asterolampra.*—Fr. single; v. circular, finely areolar, except in the centre and at equidistant clear marginal rays radiating from the centre, which is traversed by radiating dark lines (septa), alternating with the marginal rays; fossil.

69. *Asteromphalos.*—As *Asterolampra*, but two of the central dark lines parallel, and the corresponding marginal ray obliterated; fossil.

70. *Halionyx.*—Fr. single; v. circular, without septa, with rays not reaching the centre, and with intermediate shorter rays; between the rays transverse areolar lines; fossil.

71. *Odontodiscus.*—Fr. single, lenticular; v. covered with puncta (areolæ), arranged in radiating rows on excentrically curved lines, and with erect marginal teeth; fossil.

72. *Omphalopelta.*—As *Actinoptychus*, but upper part of margin of valves with a few erect spines; fossil.

73. *Symbolophora.*—Fr. single, disk-shaped; v. with incomplete septa radiating from the solid angular umbilicus, and intermediate bundles of radiating lines; marine and fossil.

74. *Systephania.*—Fr. single; v. circular, areolar, without rays or septa, with a crown of spines or an erect membrane on the outer surface of each valve; fossil.

COHORT 15. ANGULIFERA.—Valves angular.

75. *Amphitetras.*—Fr. at first united, afterwards separating into a zigzag chain, rectangular; v. rectangular, the angles often produced; marine.

76. *Amphipentras.*—Fr. solitary; v. pentangular; fossil.

77. *Lithodesmium.*—Fr. united into a straight filament; v. triangular, one side plane, the others undulate; marine.

TRIBE IV. APPENDICULATÆ.—Valves with processes or appendages, or with the angles produced or inflated.

COHORT 16. EUPODISCEÆ.—Fr. disk-shaped; v. circular.

78. *Eupodiscus.*—Fr. single, disk-shaped; v. circular, with tubular or horn-like processes on the surface; aquatic and marine.

79. *Auliscus.*—As *Eupodiscus*, but processes obtuse and more solid; fossil.

80. *Insilella.*—Fr. single, fusiform; v. equal, with a median turgid ring between them; marine.

COHORT 17. BIDDULPHIEÆ.—Fr. flattened; v. elliptical or suborbicular.

81. *Biddulphia.*—Fr. rectangular, more or less united into a continuous or zigzag filament; the angles inflated or produced into horns; v. convex, centre usually spinous; marine.

82. *Isthmia.*—Fr. rhomboidal or trapezoidal, cohering by one angle; angles produced; marine.

83. *Chætoceros.*—Fr. compressed; v. equal, with a long spine or filament on each side; marine.

84. *Rhizoselenia.*—Fr. elongate, subcylindrical, marked with transverse or spiral lines, ends oblique or conical, and with one or more terminal bristles; marine.

85. *Hemiaulus.*—Fr. single, compressed, rectangular; angles produced into tubular direct processes, those on one valve longer than on the other; fossil.

86. *Syringidium.*—Fr. single, terete, acuminate at one end, two-horned at the other; marine.

87. *Periptera.*—Fr. single, compressed; v. unequal, one simply turgid, the other with marginal wings or spines; fossil.

88. *Dicladia.*—Fr. single; v. unequal, one turgid and simple, the other two-horned; fossil.

COHORT 18. ANGULATÆ.—Valves angular.

89. *Triceratium.*—Fr. free; v. triangular, each angle with a minute tooth or horn; marine.

90. *Syndendrium.*—Fr. single, subquadrangular; v. unequal, slightly turgid, one smooth, the other with numerous median spines, or little horns branched at the ends.

B. *Frustules enveloped in a mass of Gelatin, or contained in Gelatinous Tubes, forming a Frond.*

91. *Mastogloia.*—Frond mammilate; fr. like *Navicula*, but hoops with loculi; aquatic and marine.

92. *Dickieia.*—Frond leaf-like; fr. like *Navicula* or *Stauroneis*; marine.

93. *Berkeleya.*—Frond rounded at base, filamentous at circumference; fr. navicular; marine.

94. *Homœocladia.*—Frond sparingly divided, filiform; fr. like *Nitzschia*; marine.

95. *Colletonema.*—Frond filamentous, filaments not branched; fr. like *Navicula* or *Gyrosigma*; aquatic.

96. *Schizonema.*—Frond filamentous, branched; fr. like *Navicula*; marine.

97. *Encyonema.*—Frond filamentous, but little branched; fr. like *Cymbella;* aquatic.

98. *Syncyclia.*—Fr. those of *Cymbella*, united in circular bands, immersed in an amorphous gelatinous frond; marine.

99. *Frustulia.*—Fr. as *Navicula*, irregularly scattered through an amorphous gelatinous mass; aquatic.

100. *Micromega.*—Fr. as *Navicula*, arranged in rows in gelatinous tubes, or surrounded by fibres, these being inclosed in a filiform branched frond; marine.

The family of *Nostochinæ* is allied to the *Palmellaceæ*. It consists of beaded filaments suspended in a gelatinous frond. The gelatinous masses of *Nostoc* often appear quite suddenly in damp places, and have been called "fallen stars." They attracted the notice of the alchemists, and enter into many of their recipes for the transmutation of metals. What have been termed showers of flesh or of blood, originated in all probability in the rapid development of similar masses. Many botanists regard them as the "gonidia" of *Collema* and other lichens.

The *Oscillatoria*, so called from the singular oscillatory motion of their filaments, consist also of cells which multiply in a longitudinal direction by self-division. The *Ulvaceæ*, to which the grass-green sea-weeds belong, increase in breadth as well as length by the subdivision of cells, so as to produce a leaf-like expansion (Plate X, Fig. 112). An illustration of the simpler forms of reproduction in Protophytes is seen in *Zygnema*, so called from the singular manner in which the filaments are yoked together in pairs. In an early stage of growth, while multiplication of cells proceeds by subdivision, the endochrome is generally diffused, but about the time of conjugation it arranges itself usually into a spiral. Adjacent cells put forth protuberances, which unite and form a free passage between them, and the endochrome of one cell passes over into the other and forms the spore. In *Sphæroplea* the endochrome of the "oospore" breaks up into segments, which escape as "microgonidia." Each of these have two vibratile filaments, which elongate so as to become fusiform, and at the same time change from red to green. Losing their motile power they become filaments, in which the endochrome, by the multiplication of vacuoles, becomes frothy. After a time the particles of endochrome assume a globular or ovoid shape, and openings occur in the cell-wall. In other filaments the endochrome is converted into antherozoids, each of which is furnished with two filaments, by means of which they swim about and enter the openings of the spore-cells, in which they seem to dissolve away. The contents of the spore-cell then becomes invested with a membranous envelope; the color changes from green to red; a second investment is formed within the first, which extends itself into stellate projections. When set free the mass is a true oospore, and ready to repeat the process above described. In *Œdogonium* the antherozoids are developed in a body called an "androspore," which is set free from a germ-cell, and which being furnished with cilia resembles an ordinary zoospore. This androspore attaches itself to the outer surface of a germ-cell, a sort of lid drops from its free extremity, which sets free its contained antherozoids. These enter an aperture formed in the cell-wall of the oospore, and fertilize the contained mass by blending with it.

Examination of the Higher Cryptogamia.—It would enlarge this volume far beyond its proposed limits to refer to the particular instances of form or function which the microscope reveals to the systematic botanist or physiologist, nor is this necessary, since well-written treatises on structural botany are quite available. We content ourselves, therefore, in the remainder of this chapter, with pointing out the methods of examination by which the views of other observers may be verified, or additions made to our knowledge of vegetable life.

The lower forms of algæ and fungi, to which we have already referred, need scarcely any preparation, save the disentanglement of twisted threads under the simple microscope, or a gentle teasing with needles, or rinsing with water. The solution of iodine, and of iodine and sulphuric acid, will suffice to exhibit the nature of the cell-wall and cell-contents. In more highly developed plants it will be necessary to take thin sections from different parts, and in different but definite directions. These sections may be made by hand, or between pieces of pith or cork by means of a section-cutter. In some instances some of the methods of staining will also be useful. Dr. Hunt, of Philadelphia, has proposed a plan of staining which is well adapted to all vegetable tissues. He first soaks the part or section in strong alcohol to dissolve the chlorophyll, then bleaches it in a solution of chlorinated soda. It is then placed in a solution of alum, and afterwards in one of extract of logwood. By transferring it to weak alcohol and afterwards to stronger, it is deprived of its water, and after being made transparent with oil of cloves, it is ready for mounting in balsam or dammar varnish. Care must be taken to wash it well after each of the preliminary steps before staining.

In the higher algæ, the layers of cells assume various sizes and shapes, and the nature of their fructification is of great interest. Sections may be made of the "receptacles" at the extremities of the fronds, which contain filaments, whose contents become antherozoids. The pear-shaped sporangia in the receptacles subdivide into clusters of eight cells, called octospores, which are liberated from their envelopes before fertilization.

The red sea-weeds, or *Rhodospermeæ*, afford many beautiful forms for the microscope. The "tetraspores" are imbedded in the fronds.

In *lichens*, the *apothecia* form projections from the thallus, or general expansion produced by cell-division. A vertical section shows them to contain *asci* or spore-cases amid straight filaments, or elongated cells called *paraphyses*.

The fronds of *Hepaticæ* or liverworts bear stalks with shield-like disks, which carry antheridia, and others with radiating bodies bearing archegonia, which afterwards give place to the sporangia or spore-cases. The spores are associated with *elaters*, or elastic spiral fibres, which suddenly extend themselves and disperse the spores.

The *Characeæ* are often incrusted with carbonate of lime, which may be removed with dilute sulphuric acid. The motion of the bioplasm in the cells of the stem is often well seen. The cells in which the spiral filaments or antheridia are developed, are strung together like a row of pearls. The position and construction of the spores also should be examined, as well as the mode of growth in the plant by division of the terminal cell (Plate XI, Fig. 113).

Stems of *mosses* and liverworts should be examined by means of transverse and longitudinal sections. Similar sections through the half-ripe fruit of a moss will show the construction of the fruit, the peristome, the calyptra, etc. The ripe spores may be variously examined dry, in water, in oil of lemons, and in strong sulphuric acid. The capsules or urns of mosses are not now regarded as their fructification, but its product.

The true *antheridia* and *pistillidia* are found among the bases of the leaves, close to the axis. The fertilized "embryo-cell" becomes gradually developed by cell-division into a conical body or spore-capsule, elevated on a stalk. The *peristome*, or toothed fringe, seen around the mouth of the urn when the *calyptra* or hood, and *operculum* or lid, are removed, furnishes a beautiful object for the binocular microscope.

PLATE XI.

Fig. 113.

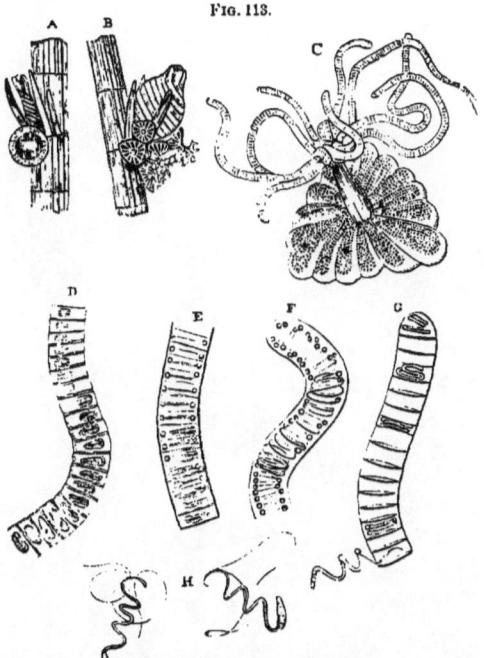

Antheridia of *Chara fragilis*:—A, antheridium or "globule" developed at the base of pistillidium or "nucule;" B, nucule enlarged, globule laid open by the separation of its valves; C, one of the valves, with its group of antheridial filaments, each composed of a linear series of cells, within every one of which an antherozoid is formed; in D, E, and F, the successive stages of this formation are seen; and at G is shown the escape of the mature antherozoids, H. (From Carpenter.)

Fig. 114.

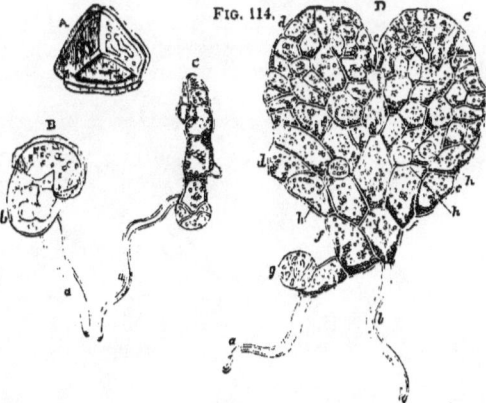

Development of Prothallium of *Pteris serrulata*:—A, spore set free from the theca; B, spore beginning to germinate, putting forth the tubular prolongation *a*, from the principal cell *b*; C, first formed linear series of cells; D, prothallium taking the form of a leaf-like expansion; *a* first and *b* second radical fibre; *c, d*, the two lobes, and *e* the indentation between them; *f, f*, first-formed part of the prothallium; *g*, external coat of the original spore; *h, h*, antheridia. (From Carpenter.)

The *Sphagnum*, or bog-moss, has large and elongated leaf-cells, with loosely-coiled spiral fibres, and their membranous walls have large apertures. Their spores are of two kinds, and when germinating in water, produce a long filament with root-fibres at the lower end and a nodule at the upper, from which the young plant is formed. If grown on wet peat, instead of a filament there is evolved a lobed foliaceous *prothallium*, resembling the frond of liverworts.

In *ferns* the structure approximates to true flowering plants, while the reproductive organs are those of cryptogamia. Thin sections of the stem, cut obliquely, show the scalariform or ladder-like vessels. The fructification is usually found on the under side of the frond in isolated spots called *sori*. Each of these contains a number of capsules or *thecæ*, and each capsule is surrounded by an *annulus* or ring, whose elasticity opens the capsule when ripe and permits the spores to escape. The spores are somewhat angular, and when vegetating give rise to a leaf-like expansion called a *prothallium*. In this the *antheridia* and *archegonia*, which represent the true flower of higher plants, are developed. The ciliated antherozoids from the antheridia penetrate the cavity of the archegonium and fertilize the "germ-cell," which subdivides and becomes a young fern, while the prothallium, having discharged the functions of a nurse, withers away (Plate XI, Fig. 114). The group of *Equisetaceæ* or horsetails is interesting from the siliceous skeletons of the epidermis, already referred to, page 131, as well as for the elastic filaments attached to their spores.

Examination of Higher Plants.

The elementary tissues described in the beginning of this chapter are chiefly characteristic of phanerogamic plants, yet some additional particulars remain to be noticed in connection with the axis or stem, the leaves, flowers, and fruit.

1. *The Stem.*—The arrangement of fibro-vascular bundles, *i. e.*, woody fibres and ducts, differs widely in the two botanical divisions of *Monocotyledons* and *Dicotyledons*. In the first the growth is *endogenous,* and a section exhibits the bundles of fibres and ducts disposed without regularity in the mass of cellular tissue which forms the basis of the fabric. In the second, or *exogenous* stems, the fibro-vascular bundles are wedge-shaped, and interposed between the bark and the pith, being kept apart by plates of cellular tissue, called medullary rays, proceeding from the pith.

The course of the vascular bundles in monocotyledons should be carefully followed, either by maceration or minute dissection. In the dicotyledonous stem, sections must be made in three directions, transversely, longitudinally across the diameter, and at a tangent from the bundles of fibres. The section-cutter, described page 63, will be serviceable, although a sharp razor or scalpel may serve. The size, form, and contents of the pith-cells should be noticed, and their transition to wood-cells. The arrangement of the medullary rays, of the wood-cells, and of the ducts must also be observed, and in the Coniferæ the position of the pits. The cambium layer, between the bark and wood, may have its cells rendered more transparent by weak alkalies, and their contents tested with iodine solution. The course and construction of laticiferous vessels in the bark, when present, and of the cork-cells of the tuberous layer, may be noted.

Fossil woods may be cut with a watch-spring saw, and ground on a hone like bone or teeth. Sometimes it is best to break off small lamella by careful strokes with a steel hammer. It is sometimes useful to digest fossil wood in a solution of carbonate of soda for several days before cutting.

2. *Leaves.*—These should be examined by thin longitudinal and transverse sections. The epidermis of both sides should be detached, and the position and arrangement of the stomata observed (Plate VII, Fig. 100). The hairs of the epidermis, the arrangement of the parenchyma, and the distribution of the vascular bundles in the form of nerves, are also of importance.

3. *Flowers.*—For ascertaining the number and position of the parts of the flower, transverse sections at different heights through an unopened bud may be taken, together with a longitudinal section exactly through the middle. The general structure of sepals and petals corresponds with that of leaves, but there are some peculiarities. Thus the cells of the petal of the geranium exhibit when deprived of epidermis, dried and mounted in balsam, a peculiar mammillated appearance with radiating hairs (Plate VIII, Fig. 102). Anthers and pollen grains are also interesting microscopic objects. The protrusion of the inner membrane through the exterior pores in pollen may be stimulated by moistening with water, dilute acid, etc. The penetration of the pollen tubes through the tissue of the style may be traced by sections or careful dissection. The heartsease, *viola tricolor*, and the black and red currant, *ribes nigrum* and *rubrum*, have been recommended for this purpose.

4. *Seeds.*—The reticulations or markings on various kinds of seeds render them frequent objects for observation with the binocular microscope. Adulterations may also be detected in this way, as well as imperfect seeds in any sample, a subject of much importance to the practical farmer.

CHAPTER XI.

THE MICROSCOPE IN ZOOLOGY.

We have already seen that both animal and vegetable structures originate in a jelly-like mass or cell, and that in the simple forms it is difficult, if not impossible, to determine whether the object is an animal or a vegetable. The mode of alimentation, and not structure, is our only guide in the discrimination of the *Protozoa* or elementary animal forms from *Protophytes* or simple vegetables.

It has been proposed by Professor Haeckel to revive the idea of a kingdom of nature intermediate between plants and animals, but it does not appear that any gain to science would result from such an arrangement.

I. Monera.—The simplest types of *Protozoa* are mere particles of living jelly (Plate XII, Fig. 115), yet they possess the power of contraction and extension, and of absorbing alimentary material into their own substance for its nutrition. The *Bathybius*, from the "globigerina mud," referred to on page 96, seems to have been an indefinite expansion of such protoplasm or bioplasm.

II. Rhizopods.—This term (meaning root-footed) is applied to such masses of sarcode or bioplasm as extend long processes, called *pseudopodia*, as prehensile or locomotive organs (Plate XII, Fig. 116). The Rhizopods are either indefinitely organized jelly, like *Monera*, or attain a covering or envelope of membrane called *ectosarc*, while the thin contents are termed *endosarc*. The first order of Rhizopods, *Reticularia*, consist of indefinite extensions of freely branching and mutually coalescing bioplasm. The second order, *Radiolaria*, have rod-like radiating extensions of the ectosarc, which do not coalesce. The order *Lobosa* are lobose extensions of the body itself, as in the

Amœba princeps already described. Some of this latter order, as *Arcella* and *Difflugia*, are testaceous. In Arcella the test is a horny membrane, analogous to the chitine which hardens the integuments of insects. In Difflugia the test is made up of minute particles of gravel, shell, etc., cemented together. From the opening the amœboid body puts forth its pseudopodia (Plate XII, Fig. 117). Connected with Rhizopods are three remarkable series of forms, generally marine, and distinguished by skeletons of greater or less density, which afford many objects of interest to the microscopist. These are the *Foraminifera*, the *Polycystina*, and the *Sponges* or *Porifera*. The shells of the Foraminifera are calcareous, and those of Polycystina siliceous; both are perforated with numerous apertures, which in Polycystina are often large. We have previously referred to these forms as occurring in a fossil state.

Some Foraminifera have porcellanous, and others vitreous or hyaline shells, usually many-chambered, and of every shape between rectilinear and spinal. Most of them are microscopic, but some are of considerable size, as the *Orbitolites*, which are found in tertiary limestones in Malabar. The Nummulitic limestone, which extends over large areas of both hemispheres, and of which the pyramids of Egypt are built, is composed of the remains of the genus *Nummulina;* and the *Eozoon Canadense* has been shown by Drs. Dawson and Carpenter to belong to the Foraminiferal type.

In some Foraminifera the true shell is replaced by a sandy envelope, whose particles are often cemented by phosphate of iron. Dr. Carpenter, whose researches have largely extended our knowledge of this group, pertinently remarks that "there is nothing more wonderful in nature than the building up of these elaborate and symmetrical structures by mere jelly specks, presenting no trace whatever of that definite 'organization' which we are accustomed to regard as necessary to the manifestations of conscious life."*

The *Polycystina*, like the Foraminifera, are beautiful objects for the binocular microscope, with the black-ground illumination by the Webster condenser, the spot-lens, or the paraboloid.

The *Porifera* or sponges begin life as solitary Amœba, and amid aggregations formed by their multiplication, the characteristic spicules of sponge-structure make their appearance. In one group, the skeleton is a siliceous framework of great beauty. In *Hyalonema*, the silica is in bundles of long threads like spun glass. Sometimes sponge spicules are needle-like, straight or curved, pointed at one or both ends; sometimes with a head like a pin, furnished with hooks, or variously stellate. Dr. Carpenter thinks it probable that each spicule was originally a segment of sarcode, which has undergone either calcification or silicification (Plate XII, Fig. 118).

III. INFUSORIAL ANIMALCULES.—From the earliest history of the microscope, the minute animals found in various infusions or in stagnant pools, etc., have attracted attention. We owe to Professor Ehrenberg the first scientific arrangement of this class, and although more extended observations have changed his classification, yet many of his views are still accepted by the most recent investigators. Ehrenberg divided this class into two groups, which represent very different grades of organization. The first he called *Polygastrica* (many-stomached) from a view of their structure, which subsequent examinations have not confirmed. The other group is that of *Rotifera* or *Rotatoria*, a form of animal life which is most appropriately classed among *worms*. The term *Infusoria* is now applied to those forms which Professor Ehrenberg

---

* The Microscope and its Revelations, by W. B. Carpenter, M.D., LL.D., etc.

Fig. 115.

Monera (*Amœba*).

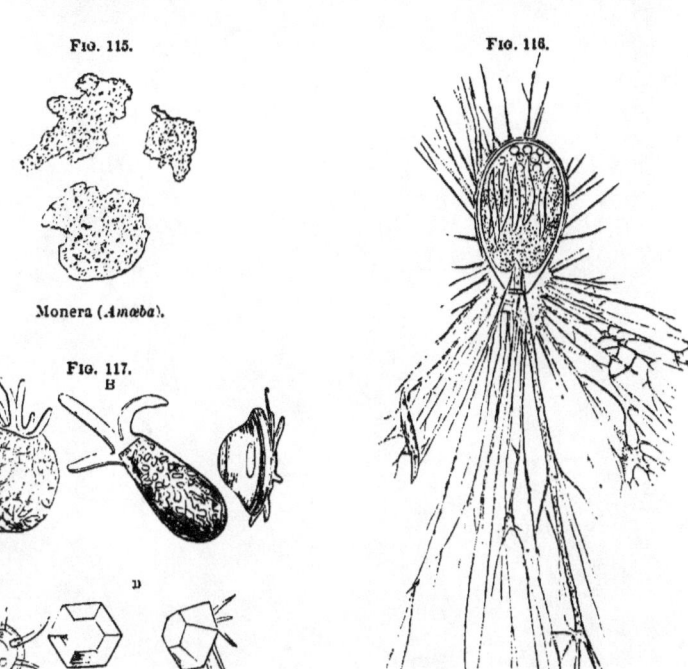

Fig. 117.

A, *Difflugia proteiformis*; B, *Difflugia oblonga*; C, *Arcella acuminata*; D, *Arcella dentata*.

Gromia oviformis, with its pseudopodia extended.

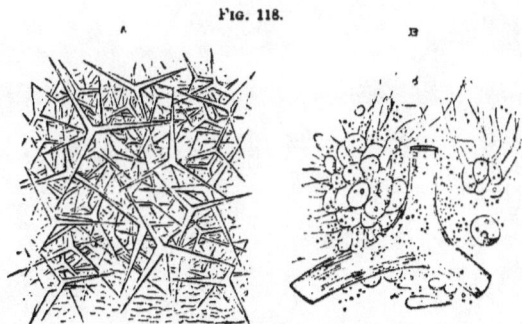

Fig. 118.

Structure of *Grantia compressa*: B, small portion highly magnified.

called polygastric animalcules. Yet a large section described by him in this connection, including the *Desmidiaceæ*, *Diatomaceæ*, *Volvocineæ*, and other protophytes, have been transferred by naturalists to the vegetable kingdom.

The bodies of the Infusoria consist of sarcode or bioplasm, having an outer layer of firmer consistence. Sometimes the integument is hardened on one side so as to form a shield, and in other cases it is so prolonged and doubled upon itself as to form a sheath or cell, within which the animalcule lies. The form of the body is more definite than that of *Amœba*, so as to be characteristic of species. It may be oblong, oval, or round; and some kinds, as *Vorticella*, are attached to a footstalk, which has the power of contracting in a spiral coil. No distinct muscular structure can be detected in the Infusoria, yet the general substance of the body is contractile. In most species short hair-like filaments or cilia project from the surface, sometimes arranged in one or more rows round the mouth, and moving to all appearance under the influence of volition. In others there are one or two flagelliform filaments, or long anterior cilia with vibratile ends. Others, again, have setæ or bristles, which assist in locomotion. The motions of some are slow, and of others quite rapid.

The interior of the sarcode body exhibit certain roundish spots, sometimes containing Diatoms or other foreign substances. They have been called gastric vesicles, cells, spaces, or sacculi. They are only visible from their contents, and seem to be mere spaces without a living membrane. If a little indigo or carmine is diffused in the water which contains the Infusoria, the cavities will soon be filled and become distinct. If watched carefully they will appear to move round the body of the animal, and as the pigment escapes at some part of the surface, the spots will disappear. Ehrenberg regarded these spots as so many stomachs arranged about a common duct, but the common opinion at present regards them as temporary digestive sacs made by the inclosure of food by the soft bioplasm.

In addition to the "vacuoles" described, contractile vesicles are seen which contract and dilate rhythmically, and do not change their position. They have been considered to serve for respiration.

Most of the Infusoria multiply by self-division (Plate XIII, Fig. 119), and at certain times undergo an encysting process, much resembling the "still" condition of Protophytes, and like that serving for preservation under circumstances which are unfavorable to ordinary vital activity. The gemmules or progeny which result from the bursting of the cyst do not always resemble the parent in form. The recent researches of Drs. Dallinger and Drysdale have shown considerable variety in the life history of the Infusoria. In some instances the product of the encysting process was not a mass of granules, but an aggregation of minute germinal particles not more than $\frac{1}{200000}$th of an inch in diameter, and capable of resisting heat, either by boiling or by dry heating up to 300° F.

The observations of M. Balbiani show that in many of the Infusoria, male and female organs are combined in the same individual, but that a congress of two is necessary for the impregnation of the ova, those of each being fertilized by the spermatozoa of the other.

There is also a curious tribe of suctorial animalcules called *Acinetæ*, which put forth tubular prolongations which penetrate the bodies of other species and grow in their interior as parasites.

The systematic arrangement of the Infusoria is yet unsettled. Ehrenberg's families, excluding those now placed among Algæ or Rhizopods, are as follows:

A. Intestinal tube absent.
   Body variable, without cilia.
      Carapace absent, ... ASTASIÆA.
      Carapace present, ... DINOBRYINA.
   Cilia or setæ present.
      Carapace absent, . . . CYCLIDINA.
      Carapace present, . . . PERIDINÆA.
B. Intestinal tube present.
   Orifice single.
      Carapace absent, . . . VORTICELLINA.
      Carapace present, . . . OPHRYDINA.
   Two opposite orifices.
      Carapace absent, . . . ENCHELIA.
      Carapace present, . . . COLEPINA.
   Orifices differently placed.
      Carapace none.
         No tail, but a proboscis, . . TRACHELINA.
         Tail present, mouth anterior, . OPHRYOCERCINA.
      Carapace present, ... ASPIDISCINA.
   Orifices ventral.
      Carapace absent.
         Motion by cilia, . . . COLPODEA.
         Motion by organs, . . . OXYTRICHINA.
      Carapace present, ... EUPLOTA.

IV. ROTATORIA OR WHEEL ANIMALCULES.—These are microscopic, aquatic, transparent animals, of a higher organization than the Infusoria, and belonging in all probability to the class *Vermes*. Their chief interest to the microscopist is derived from the possession of a more or less lobed, retractile disk, covered with cilia, which, when in motion, resemble revolving wheels. They have also a complicated dental apparatus, and generally a distinct alimentary canal, and are reproduced by ova. Some are more or less covered by a carapace, and in most there is a retractile tail-like foot, sometimes terminated by a suctorial disk or a pair of claw-like processes. The nervous and vascular systems are not well known, although traces of them are seen. The young of some possess an eye which often disappears in the adult. They are remarkably tenacious of life, having in some instances revived after having been kept dry for several years.

M. Dujardin divides the Rotifera into four groups or natural families:

1. Those attached by the foot, which is prolonged into a pedicle. It includes two families, the *Floscularians* and the *Melicertians*, in the first of which the sheath or carapace is transparent, and in the other composed of little rounded pellets (Plate XIII, Fig. 120).

2. The common *Rotifer* and its allies, which swim freely or attach themselves by the foot at will (Plate XIII, Fig. 121).

3. Those which are seldom or never attached, the *Brachionians* and the *Furcularians*. The former are short, broad, and flat, and inclosed in a sort of cuirass; the latter are named from a bifurcated, forcep-like foot (Plate XIV, Fig. 122).

4. The *Tardigrada* or water bears. These have no ciliated lobes, but are in other respects like their allies, and seem to be a connecting link between the Rotifers and worms. The segments of the body, except the head, bear two fleshy protuberances furnished with four curved hooks.*

V. POLYPS.—The animals of this class were formerly called *Zoophytes*, or animal flowers. They are the most important of coral-making animals, although the Hydroids and Bryozoa, together with some Algæ, as the Nullipores, share with them the formation of coral, which is a secretion of calcareous matter. Dana's work on corals gives a classification, of which we present a summary.

A good idea of a polyp may be had from comparison with the garden aster, the most common form of a polyp flower being a disk fringed with petal-like organs called tentacles.

The internal structure, like the external, is radiate, and

---

* Carpenter on the Microscope.

PLATE XIII.

FIG. 119.

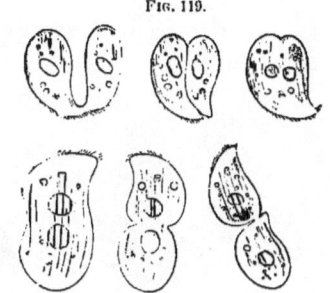

Fissiparous multiplication of *Chilodon cucullulus*.

FIG. 120.

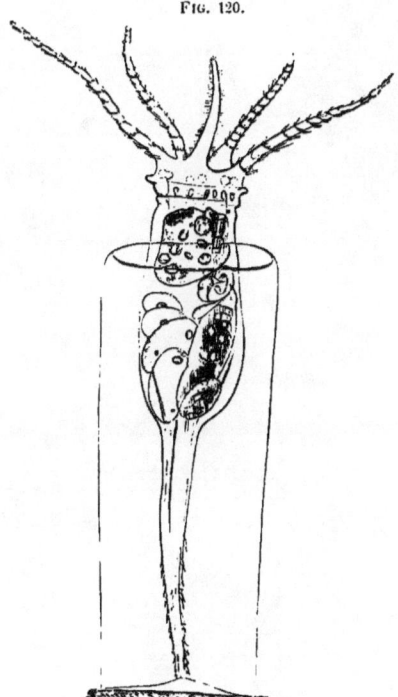

Stephanoceros Eichornii.

FIG. 121.

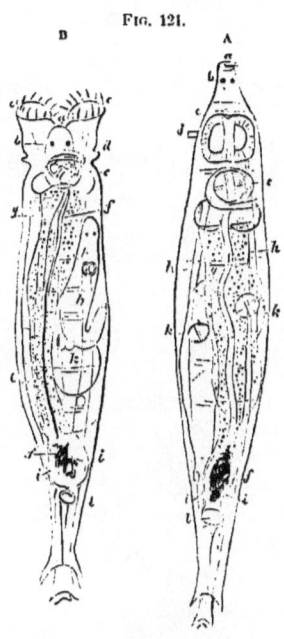

Rodifer vulgaris, as seen at A, with the wheels drawn in, and at B with the wheels expanded; *a*, mouth; *b*, eye-spots; *c*, wheels; *d*, calcar (antenna?); *e*, jaws and teeth; *f*, alimentary canal; *g*, glandular (?) mass enclosing it; *h*, longitudinal muscles; *i, i*, tubes of water vascular system; *k*, young animal; *l*, cloaca.

the cavity of the body is divided by septa into narrow compartments. The walls contain circular and longitudinal muscles, which serve for contraction of the body, which is afterwards expanded by an injection or absorption of water by the mouth.

The most interesting part of the structure of these animals, to the microscopist, is the multitude of *lasso-cells*, called also *nettling-cells, thread capsules*, and *cnidæ*, which stud the tentacles and other parts of the body, and by means of which the prey of the polyp is at once pierced and poisoned. A small piece of the tentacle of a sea anemone placed in a compressorium under the microscope, and subjected to gentle pressure, will show the protrusion of many little dart-like processes attached to thread-like filaments. Many observations indicate the injection of a poison through these darts, which is instantly fatal to small animals (Plate XIV, Fig. 123).

The polyp has no circulating fluid but the results of digestion mixed with salt water, no bloodvessels but the vacuities among the tissues, and no passage for excrements except the mouth and the pores of the body. Reproduction is both by ova and by buds.

I. *Actinoid polyps* are related to the Actinea or sea anemone. The number of tentacles and interior septa is a multiple of six.

II. *Cyathophylloid polyps* have the number of tentacles and septa a multiple of four.

III. *Alcyonoid polyps* have eight fringed tentacles. The Alcyonium tribe are among the most beautiful of coral shrubs. The Gorgonia tribe has reticulated species like the sea fan, and bears minute calcareous spicules, often brilliantly colored. The Pennatula tribe is unattached, and often rod-like, with the polyps variously arranged.

VI. HYDROIDS.—The type of this class is the common Hydra, which is often found attached to leaves or stems of aquatic plants, etc. It is seldom over half an inch long.

It has the form of a polyp, with long slender tentacles. Besides these tentacles with their lasso-cells, it has no special organs except a mouth and tubular stomach. Like the fabled Hydra, if its head be cut off another will grow out, and each fragment will in a short time become a perfect animal, supplying whatever is wanting, hence its name (Plate XIV, Fig. 124). The Hydra has the power of locomotion, bending over and attaching its head until the tail is brought forward, somewhat after the manner of a leech.

Compound Hydroids may be likened to a Hydra whose buds remain attached and develop other buds until an arborescent structure, called a *polypary*, is produced. The stem and branches consist of fleshy tubes with two layers, the inner one having nutritive functions, and the outer secreting a hard, calcareous, or horny layer. The individuals of the colony are of two kinds, the *polypite* or nutritive zooid, resembling the Hydra, and the *gonozooid*, or sexual zooid, developed at certain seasons in buds of particular shape.

To mount compound Hydrozoa, or similar structures, place the specimen alive in a cell, and add alcohol drop by drop to the sea-water; this will cause the animals to protrude and render their tentacles rigid. Then replace the alcohol with Goadby's solution, dilute glycerin, or other preserving fluid.

VII. ACALEPHS, or *sea-nettles*, are of all sizes, from an almost invisible speck to a yard in diameter. They swarm in almost every sea, and are frequently cast upon the beach by the waves. They are transparent, floating free, discoid or spheroid, often shaped like a mushroom or umbrella, and their organs are arranged radiately round an axis occupied by the pedicle or stalk. They are furnished with muscular, digestive, vascular, and nervous systems. They were formerly divided into

1. *Pulmonigrada*, from their movements being effected

PLATE XIV.

Fig. 122.

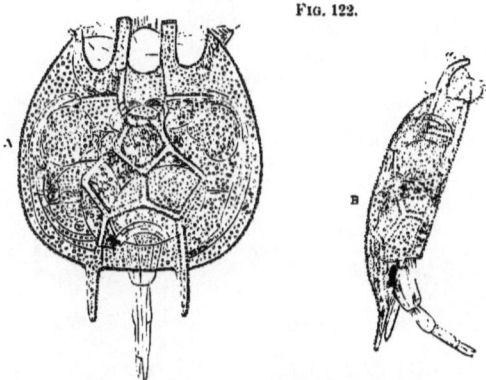

Noteus quadricornis:—A, dorsal view; B, side view.

Fig. 123.

Fig. 124.

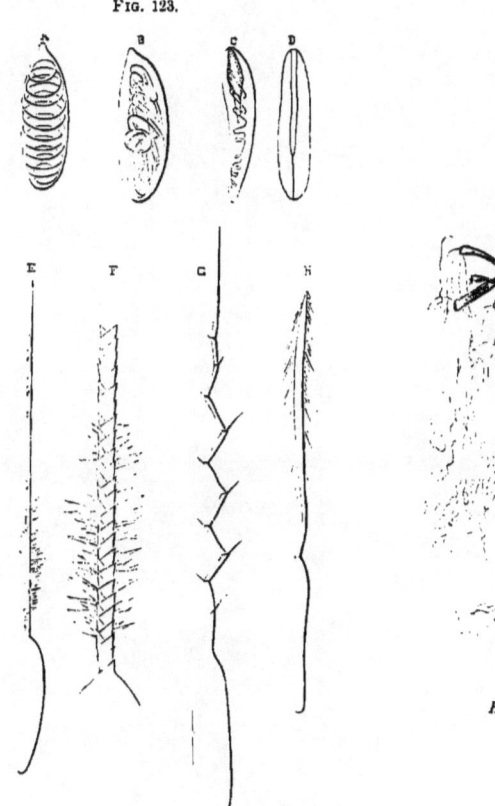

Hydra fusca in gemmation.

Filiferous capsules of Helianthoid Polypes:—A, B, *Corynactis Allmanni;* C, E, F, *Caryophyllia Smithii;* D, G, *Actinia crassicornis;* H, *Actinia candida.*

by a rhythmical contraction and dilation, as in *Rhizistoma*, etc. 2. *Cilograda*, moving by narrow bands of vibratile cilia variously disposed over the body. In *Beroe* the cilia are transformed into flat fin-like shutters, arranged in eight longitudinal bands. In Venus's girdle, *Cestum Veneris*, the margins of a gelatinous ribbon are fringed with cilia. 3. *Physograda*, which move by means of an expansile bladder, as the *Physalia*, or Portuguese man of war. 4. *Cirrigrada*, possessing a sort of cartilaginous skeleton, and furnished with appendages called cirri, serving as oars and for prehension, as *Porpita* and *Velella*. In the latter there is also a subcartilaginous plate rising at right angles from the surface supporting a delicate membrane, which acts as a sail.

This classification has been laid aside since the microscopic discovery of the close relationship between the Hydrozoa and the Medusoid Acalephs, and the latter are now subdivided into the "naked-eyed" and the "covered-eyed" Acalephs. The alternation of generations, page 126, is fully illustrated in this class. The embryo emerges as a ciliated gemmule, resembling one of the Infusoria. One end contracts and attaches itself so as to form a foot, while the other enlarges and becomes a mouth, from which four tubercles sprout and become tentacles. Thus a Hydra-like polyp is formed, which acquires additional tentacles. From such a polyp many colonies may rise by gemmation or budding, but after a time the polyp becomes elongated, and constricted below the mouth. The constricted part gives origin to other tentacles, while similar constrictions are repeated round the lower parts of the body, so as to divide it into a series of saucer-like disks, which are successively detached and become Medusæ (Plate XV, Figs. 125, 126).

VIII. ECHINODERMS.—This class includes the star-fishes, the sea-urchins or sea-eggs, the sea-slugs, and the crinoids or stone lilies of former ages. If we imagine a polyp with a long stem to secrete calcareous matter, not merely externally, but in the substance of its body and tentacles, such polyp when dried would present some such appearance as the fossil Encrinoid Echinoderms of past times. The imagination of such a polyp without a stem, and having sucker-like disks on its arms, will give us the picture of a star-fish (*Asterias*). Imagine the rays diminished and the central part extended, either flat or globular, and we have the form of *Echini* with the spines removed. The *Holothuriæ* have elongated membranous bodies, with imbedded spiculæ.

The structure of Echinoderms is quite complex, and belongs to comparative anatomy rather than microscopy, yet some directions for the study of these forms is essential to our plan.

Thin sections of the shells, spines, etc., may be made by first cutting with a fine saw, and rubbing down with a flat file. They should be smoothed by rubbing on a hone with water, cemented to a glass slip with balsam, and carefully ground down to the required thickness. They may be mounted in fluid balsam.

Many Echinoderms have a sort of internal skeleton formed of detached plates or spiculæ. The membranous integument of the Holothuriæ have imbedded calcareous plates with a reticulated structure, and they are often furnished with appendages, as prickles, spines, hooks, etc., which form beautiful microscopic objects.

The larva of an Echinoderm is a peculiar zooid, which develops by a sort of internal gemmation. One of the most remarkable of these larvæ has been called *Bipinnaria*.

IX. BRYOZOA OR POLYZOA.—Microscopic research has removed this class from the polyps, which they resemble, to the molluscan sub-kingdom. They have a group of ciliated tentacles round the mouth, but have a digestive system far more complex than polyps. They form delicate

PLATE XV.

Fig. 125.

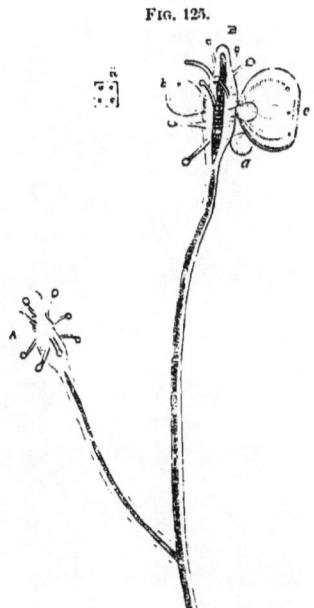

Development of Medusa buds in *Synchoryna Sarsii*.

Fig. 127.

A, Portion of *Cellularia ciliata*, enlarged; B, one of the "bird's-head" of *Bugula avicularia*, more highly magnified, and seen in the act of grasping another.

Fig. 126.

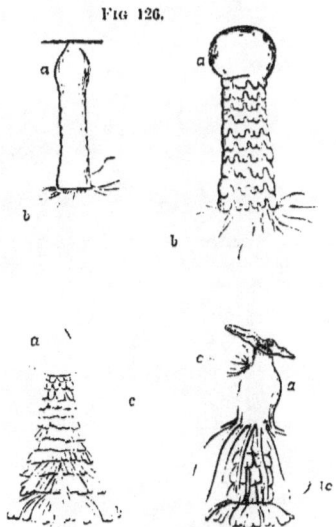

Successive stages of development of Medusa buds from *Strobila larva*.

Fig. 128.

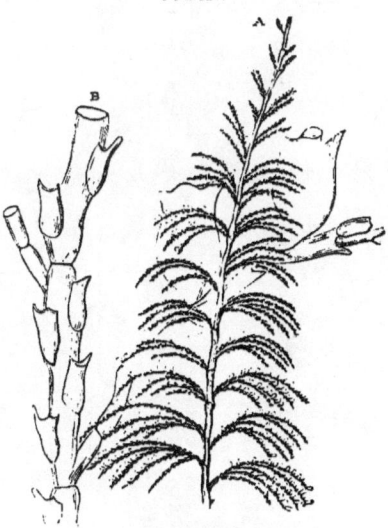

Sertularia cupressina:—A, natural size; B, portion magnified.

corals. either membranous or calcareous, made up of minute cabin-like cells, which are either thin crusts on sea-weeds, rocks. etc., or slender moss-like tufts, or groups of thin curving plates, or net-like fronds, and sometimes thread-like lines or open reticulations. The cells of a group have no connection with a common tube, as the Hydroids, but the alimentary system of each little Bryozoon is independent.

Many of the Polyzoa have curious appendages to their cells, of two kinds; the first are called birds'-head processes or *avicularia*. They consist of a body, a hinge or lower jaw-like process, and a stalk. The lower portion is moved by an elevator and depressor muscle, and during life the motion is constant. The second kind, or *vibracula*, is a hollow process from which vibratile filaments project (Plate XV, Figs. 127, 128).

X. TUNICATA.—These molluscs are so named from the leathery or cartilaginous tunic which envelops them, and which often contains calcareous spicula. Like the Bryozoa they tend to produce composite structures by gemmation, but they have no ciliated tentacles. They are of most interest to the microscopist from the peculiar actions of their respiratory and circulatory organs, which may be seen through the transparent walls of small specimens. The branchial or respiratory sac has a beautiful network of bloodvessels, and is studded with vibratile cilia for diffusing water over the membrane. The circulation is remarkable from the alternation of its direction.

The smaller Tunicata are usually found aggregate, investing rocks, stones and shells, or sea-weeds; a few are free.

Synopsis of the Families.

A. Attached; mantle and test united only at the orifices.

1. *Botryllidæ.*—Bodies united into systems.

2. *Clavelinidæ.*—Bodies distinct, but connected by a common root thread.

3. *Ascidiadæ.*—Bodies unconnected.

B. Free; mantle and test united throughout.

4. *Pelonææ.*—Orifices near together.

5. *Salpadæ.*—Orifices at opposite ends.

XI. CONCHIFERA.—This class consists of bivalve molluscs, and is chiefly interesting to the microscopist from the ciliary motion on their gills and the structure of the shell.

The ciliary motion may be observed in the oyster or mussel, by detaching a small piece of one of the bands which run parallel with the edge of the open shell, placing it on a glass slide in a drop of the liquid from the shell, separating the bars with needles, and covering it with thin glass: or the fragment may be placed in the live box and submitted to pressure. The peculiar movement of each cilium requires a high magnifying power. It appears to serve the double purpose of aeration of the blood and the production of a current for the supply of aliment.

Dr. Carpenter has shown that the shells of molluscs possess definite structure. In the *Margaritaceæ* the external layer is prismatic, and the internal nacreous. The nacreous or iridescent lustre depends on a series of grooved lines produced by laminæ more or less oblique to the plane of the surface. The shells of *Terebratulæ* are marked by perforations, which pass from one surface to another. The rudimentary shell of the cuttle-fish (of the class *Cephalopoda*), or "cuttle-fish bone," is a beautiful object either opaque or in the polariscope. Sections may be made in various directions with a sharp knife, and mounted as opaque objects or in balsam.

XII. GASTEROPODA.—These molluscs are either naked, as the slug, or have univalve shells, as the snail, the limpet, or the whelk. As in the other classes referred to, the details of anatomical structure are full of interest;

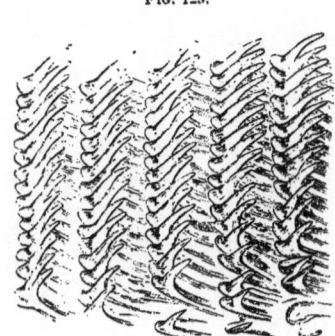

Fig. 129.

Palate of *Doris tuberculata*.

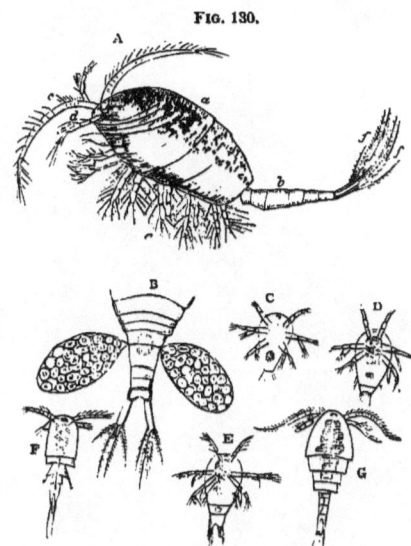

Fig. 130.

A, female of *Cyclops quadricornis*;—*a*, body; *b*, tail; *c*, antenna; *d*, antennule; *e*, feet; *f*, plumose setæ of tail;—B, tail, with external egg-sacs; C, D, E, F, G, successive stages of development of young.

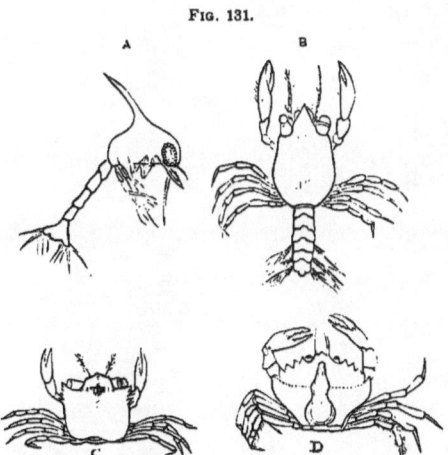

Fig. 131.

Metamorphosis of *Carcinus mænas*:—A, first stage; B, second stage; C, third stage, in which it begins to assume the adult form; D, perfect form.

but to the microscopist the palate, or tongue as it is called —a tube which passes beneath the mouth, opening obliquely in front, and which is covered with transverse rows of minute teeth set upon plates—presents characters of great value in classification. These palates require careful dissection, and when mounted in balsam become beautiful polariscope objects (Plate XVI, Fig. 129).

XIII. CEPHALOPODA.—The crystalline lens in the eye of the cuttle-fish is said to be of the same form as the well-known "Coddington lens." The skin of this class contains a curious provision for changing its hue, consisting of large pigment-cells containing coloring matter of various tints.

The suckers, or prehensile disks, on the arms of cephalopods often make interesting opaque objects when dried.

XIV. ENTOZOA.—These are parasitic animals belonging to the class of worms. They are characterized by the absence or low development of the nutritive system, and the extraordinary development of their reproductive organs. Thus the *Tænia* or tapeworm has neither mouth nor stomach, the so-called "head" being merely an organ for attachment, while each segment of the "body" contains repetitions of a complex generative apparatus.

Among the Nematoid or roundworms, the *Anguillulæ*, or little eel-like worms, found in sour paste, vinegar, etc., as well as the *Trichina spiralis*, inhabiting the voluntary muscles, are generally classified.

ORDER I. STERELMINTHA.—Alimentary canal absent or indistinct.

FAMILY 1. *Cestoidea.*—Tapeworms; body strap-shaped, divided into transverse joints; alimentary canal indistinct. The cystic Entozoa (*Echinococcus*, etc.) are nurse or larval forms of *Cestoidea*.

FAMILY 2. *Trematoda.*—Body mostly flattened; alimentary canal distinct; branched.

FAMILY 3. *Acanthocephala.*—Body flattened, transversely wrinkled; sexual organs in separate individuals.

FAMILY 4. *Gordiacea* (Hairworms).—Body filamentous, cylindrical; alimentary canal present; sexes distinct.

FAMILY 5. *Protozoidea* or *Gregarinida.*—Probably larval forms.

ORDER II. CÆLELMINTHA.—Alimentary canal distinct.

FAMILY 1. *Nematoidea* (Roundworms).—Body cylindrical, hollow; sexes separate.

The *Enoplidæ* tribe is distinguished by an armature of hooks or styles round the mouth. Most of them are microscopic.

XV. ANNULATA (Red-blooded Worms).—Some of these, as the *Serpula*, etc., are inclosed in tubes formed of a shelly secretion, or built up of grains of sand, etc., agglutinated together. Many have special respiratory appendages to their heads, in which the microscope will exhibit the circulation. The worms of the *Nais* tribe, also, are so transparent as to be peculiarly fitted for microscopic study of structure. The dental apparatus of the leech consists of a triangular aperture in a sucking disk, furnished with three semicircular horny plates, each bordered with a row of eighty to ninety teeth, which act like a saw.

ORDER 1. TURBELLARIA.—Body bilateral, soft, covered with vibratile cilia, not segmented; eyes distinct; sexless or hermaphrodite.

ORDER 2. SUCTORIA (Apoda).—Body elongate, ringed, without bristles or foot-like tubercles; locomotion by sucking-disks; no external branchiæ.

ORDER 3. SETIGRADA (Chætopoda).—Body ringed, elongate, with feet or setigerous rudiments of them; external branchiæ usually present.

XVI. CRUSTACEA.—In the family of *Isopoda* the microscopist will find the *Asellus vulgaris*, or water wood-louse, of great interest, as readily exhibiting the dorsal vessel and circulating fluids.

The family of *Entomostraca* contains a number of genera, nearly all of which are but just visible to the naked eye. They are distinguished by the inclosure of the body in a horny or shelly case, often resembling a bivalve shell, though sometimes of a single piece. The tribe of *Lophyropoda* (bristly-footed), or "water-fleas," is divided into two orders, the first of which, *Ostracoda*, is characterized by a bivalve shell, a small number of legs, and the absence of an external ovary. A familiar member of this order, the little *Cypris*, is common in pools and streams, and may be recognized by its two pairs of antennæ, the first of which is jointed and tufted, while the second is directed downwards like legs. It has two pairs of legs, the posterior of which do not appear outside the shell.

The order *Copepoda* has a jointed shell, like a buckler, almost inclosing the head and thorax. To this belongs the genus *Cyclops* (named from its single eye), the female of which carries on either side of the abdomen an egg capsule, or external ovarium, in which the ova undergo their earlier stages of development (Plate XVI, Fig. 130).

The *Daphnia pulex*, or arborescent water-flea, belongs to the order *Cladocera* and tribe *Branchiopoda*. The other order of this tribe, the *Phyllopoda*, has the body divided into segments, furnished with leaf-like members or "fin feet."

When first hatched, the larval Entomostraca differ greatly from the adult. The larval forms of higher Crustacea resemble adult Entomostraca.

The suctorial Crustacea, order *Siphonostoma*, are generally parasitic, mostly affixed to the gills of fishes by means of hooks, arms, or suckers, arising from or consisting of modified foot-jaws. The transformations in this order, as in the *Lernæa*, seem to be a process of degradation. The young comes from the egg as active as the young of *Cyclops*, which they resemble, and pass through a series of metamorphoses in which they cast off their locomotive members and their eyes. The males and females do not resemble each other.

The order *Cirrhipeda* consists of the *barnacles* and their allies. In their early state they resemble the Entomostraca, are unattached, and have eyes. After a series of metamorphoses they become covered with a bivalve shell, which is thrown off; the animal then attaches itself by its head, which in the barnacle becomes an elongated pedicle, and in Balanus expands into a disk. The first thoracic segment produces the "multivalve" shell, while the other segments evolve the six pairs of cirrhi, which are slender, tendril-like appendages, fringed with ciliated filaments.

In the order *Amphipoda*, the *Gammarus pulex*, or fresh-water shrimp, and the *Talitrus saltator*, or sandhopper, may be interesting to the microscopist.

The order *Decapoda*, to which belong the crab, lobster, shrimp, etc., is of interest, from the structure of the shell and the phenomena of metamorphosis. The shell usually consists of a horny structureless layer exteriorly, an areolated stratum, and a laminated tubular substance. The difference between the adult and larval forms in this order is so great that the young crab was formerly considered a distinct genus, *Zoea* (Plate XVI, Fig. 131).

For the preservation of specimens of Crustacea, Dr. Carpenter recommends glycerin jelly as the best medium.

XVII. INSECTS.—Many insects may be mounted dry, as opaque objects. They may be arranged in position by the use of hot water or steam. Those which are transparent enough may be mounted in balsam, and very delicate ones in fluid. To display the external chitinous covering of an entire insect, it may be soaked in strong liquor potassæ, and the internal parts squeezed out in a saucer of water by gently rolling over it a camel's-hair brush. It may be put on a slide, and the cover fastened by tying with a thread. It should then be soaked in turpentine until quite transparent, when it may be removed, the turpentine partially drained off, and a solution of balsam in chloroform allowed to insinuate itself by capillary attraction. Gentle heat from a spirit-lamp will be useful at this stage of the mounting.

Small insects hardly need soaking in caustic potash, as turpentine or oil of cloves will render them after awhile quite transparent, and their internal organs are beautifully seen in the binocular microscope. Thin sections of insects are instructive, and may be made with a section-cutter by first saturating the body with thick gum mucilage, and then incasing in melted paraffin

Many insects and insect preparations are well preserved in glycerin.

The eggs of insects are often interesting objects, and should be mounted in fluid.

Wing cases of beetles are often very brilliant opaque objects. Some are covered with iridescent scales, and others have branching hairs. Many are improved by balsam. and this may be determined by touching with turpentine before mounting.

Scales of Lepidoptera, etc., may be exhibited in their natural arrangement by mounting a small piece of wing dry. If desired as test objects, a slide or thin cover, after having been breathed on, may be slightly pressed on the wing or body of the insect. The scales are really flattened cells, analogous to the epidermic cells of higher animals. Some have their walls strengthened by longitudinal ribs, while others, as the *Podura*, show a beaded appearance under high powers from corrugation. Dr. Carpenter believes the exclamation marks in the scales of the latter to be the most valuable test of the excellence of an objective.

Hairs of insects are often branched or tufted. The hair of the bee shows prismatic colors if the chromatic aberration of the object-glass is not exactly neutralized.

Antennæ vary greatly in form, and are often useful in

classification (Plate XVII, Fig. 132). Thus in the *Coleoptera* we have the *Serricornes*, or serrated antennæ; the *Clavicornes*, or clubbed; the *Palpicornes*, with antennæ no larger than palpi; the *Lamellicornes*, with leaf-like appendages to the antennæ; and the *Longicornes*, with antennæ as long or longer than the body. Nerve-fibres, ending in minute cavities in the antennæ, have been traced, which are supposed to be organs of hearing. The antennæ should be bleached to exhibit them. The bleaching process is also useful for other parts of insects. The bleaching fluid consists of a drachm of chlorate of potass in about two drachms of water, to which is added about a drachm of hydrochloric acid.

Compound eyes of insects are always interesting. They are quite conspicuous, and often contain thousands of facets, or minute eyes, called *ocelli* (Plate XVII, A B, Fig. 133). Besides these, insects possess rudimentary single eyes, like those of the *Arachnidæ*. These are at the top of the head, and are termed *stemmata* (Plate XVII, *a*, Fig. 133). To display the "corneules," or exterior layer of the compound eye, the pigment must be carefully brushed away after maceration. A number of notches may then be made around the edge, the membrane flattened on a slide, and mounted in balsam. Vertical sections may be made while fresh, so as to trace the relations of the optic nerve, etc. The dissecting microscope and needles will be found useful (Plate XVII, Fig. 132).

Mouths of insects present great varieties. In the beetles the mouth consists of a pair of mandibles, opening laterally; a second pair, called maxillæ; a labrum or upper lip; an under lip or labium; one or two pairs of jointed appendages to the maxillæ, termed maxillary palpi; and a pair of labial palpi. The labium is often composed of distinct parts, the first of which is called the mentum or chin, and the anterior part the ligula or tongue. This latter part is greatly developed in the fly, and presents

PLATE XVII.

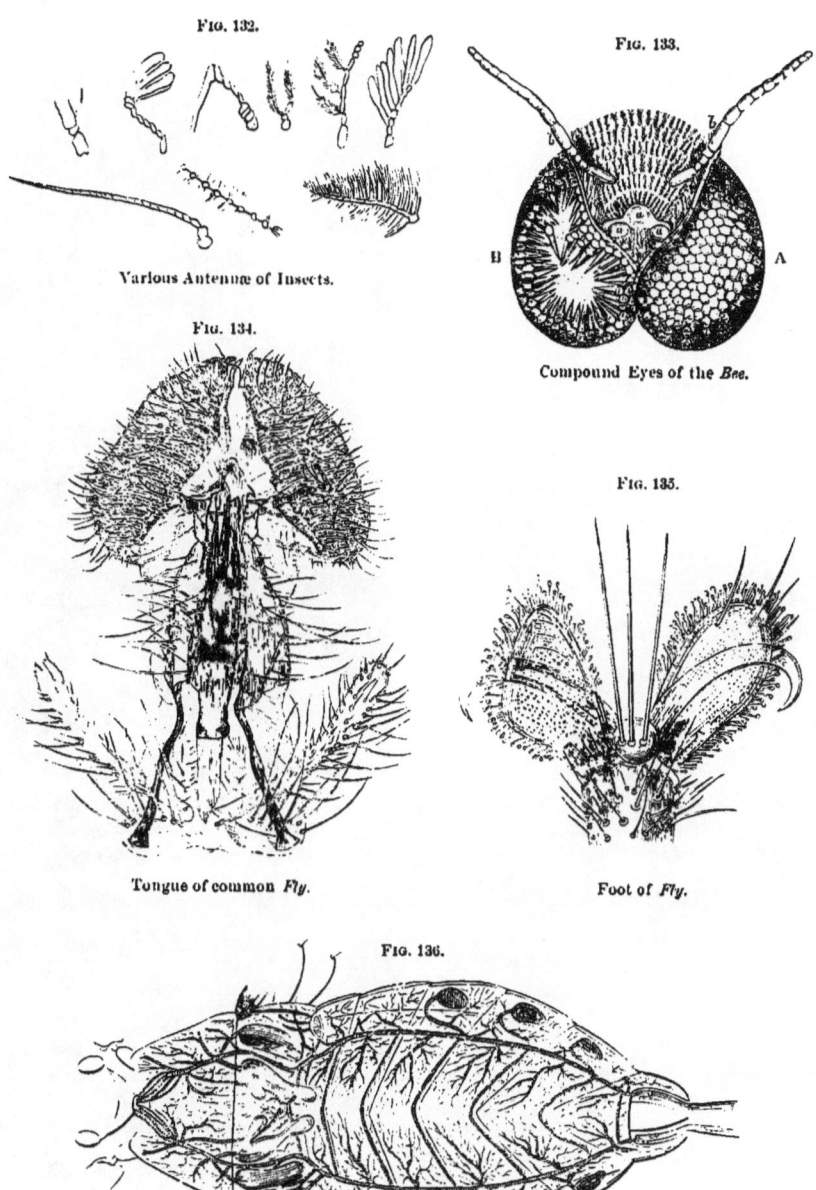

Fig. 132.

Various Antennæ of Insects.

Fig. 133.

Compound Eyes of the *Bee*.

Fig. 134.

Tongue of common *Fly*.

Fig. 135.

Foot of *Fly*.

Fig. 136.

Tracheal system of *Nepa* (Water-scorpion).

a curious modification of tracheal structure, which is thought to serve the function of suction (Plate XVII, Fig. 134). The tongue of the bee is also an interesting object. In the *Diptera* the labrum, maxillæ, mandibles, etc., are converted into delicate lancets, termed setæ, and are used to puncture the epidermis of animals or plants, from which the juices may be drawn by the proboscis. In the *Lepidoptera* the labrum and mandibles are reduced to minute plates, while the maxillæ are greatly elongated, and are united to form the haustellum, or true proboscis, which contains a tube for suction.

Feet.—These organs vary with the habits of life in different species. The limb consists of five divisions: the coxa or hip, the trochanter, the femur or thigh, the tibia or shank, and the tarsus or foot. This last has usually five joints, but sometimes less. The *Coleoptera* are subdivided into groups, according as the tarsus consists of five, four, or three segments. The last joint is furnished with hooks or claws, and in the fly, etc., the foot is also furnished with membranous expansions, called pulvilli. These latter have numerous hairs, each of which has a minute disk at its extremity. By these, probably by the secretion of a viscid material, the insect is enabled to walk on glass, etc., in opposition to gravity (Plate XVII, Fig. 135). In the *Dytiscus*, the inner side of the leg is furnished with disks or suckers of considerable size. They may be mounted as opaque objects. *Stings* and *Ovipositors* also present a great variety of structure, and may be best mounted in balsam.

The *alimentary canal* in insects presents many diversities. As in higher animals, it is shorter in flesh-eaters than in feeders on vegetables. It consists of: 1. The œsophagus, which is sometimes dilated to form a crop. 2. The muscular stomach, or gizzard, whose lining membrane is covered with plates, or teeth, for trituration. 3. A cylindrical true stomach, in which digestion takes place. 4. The small intestine, terminating in 5, the large intestine or colon. The colon of most insects in the imago or perfect state, never in larvæ or pupæ, contains from four to six organs of doubtful nature arranged in pairs. They are transparent, round, or oval tubercles projecting inside the colon, traversed by tufts of tracheæ, and sometimes with a horny ring at the base.

The salivary glands are sacs or tubes of variable form and length, terminating near the mouth. A distinct *liver* is absent, its function being performed by glandular cells in the walls of the stomach. Many insects, however, have cæcal appendages to the stomach which secrete bile. Some have tubular cæca appended to the small intestine, probably representing a *pancreas*. In the interspaces of the various abdominal organs, is found a curious organ called the fatty body, which attains its development at the close of the larval period, and appears to form a reservoir of nourishment for the pupa. It consists of fat-cells imbedded in a reticular tissue, and is traversed by slender tracheæ.

The *Malpighian vessels* are slender, mostly tubular glands, cæcal or uniting with each other, which open into the pyloric end of the stomach, and as uric acid has been found in them, are thought to serve the functions of a *kidney*. Some consider the renal organ to be represented by certain long vessels convoluted on the colon, and opening near the anus.

Other glandular organs occur in insects, as cysts in the integument, called glandulæ odoriferæ; poison glands, attached to the sting in many females; and silk-secreting glands, coiled in the sides of the body and opening outside the mouth.

The *heart* is a long contractile vessel situated in the back. It is constricted at intervals. The posterior part acts as a heart, and the anterior represents an aorta, and conveys blood to the body. From the anterior end the blood passes in currents in all directions, without vascular walls, running into the antennæ, wings, extremities, etc., and returning as a venous current, forming two lateral currents towards the end of the abdomen, it is brought by the diastole of the heart through lateral fissures existing in it.

The respiration is effected by means of *tracheæ*, two or more large vessels running longitudinally, giving off branches in all directions, and opening to the air by short tubes, connected at the sides of the body with orifices called *spiracles*. Aquatic larvæ often have branchiæ in the form of plates, leaves, or hairs, through which the tracheæ ramify (Plate XVII, Fig. 136).

The *nervous system* consists of a series of ganglia arranged in pairs, one for each segment of the body. They are situated between the alimentary canal and the under surface of the body, and are usually connected by longitudinal nervous cords. From the ganglia nerves are distributed to all parts.

The *muscular system* of insects is quite extensive. Lyonet dissected and described more than four thousand in the caterpillar of the goat-moth (*Cossus ligniperda*).

XVIII. ARACHNIDA.—This class of animals includes mites, ticks, spiders, and scorpions. They are destitute of antennæ; the head and thorax are united; they have simple eyes (ocelli), and eight jointed legs.

The cheese-mite, the "ticks," the itch-insect (*Sarcoptes scabies*), and the *Demodex folliculorum*, which is parasitic in the sebaceous follicles of the skin of the face, are common examples of *Acari*. They are best mounted in fluid.

The respiratory apparatus in spiders differs from that of insects, the spiracles opening into respiratory sacs, which contain leaf-like folds for aeration of blood. The spinning apparatus is also interesting.

The minute anatomy of vertebrated animals affords the

microscopist numerous specimens, but the details will be best understood from the following chapter.

As the classification of the Invertebrata is subject to great variation, the following table, after Nicholson, is added for the sake of comparison:

INVERTEBRATE ANIMALS.

SUB-KINGDOM I.—PROTOZOA.

CLASS I. GREGARINIDÆ.—Parasitic Protozoa, destitute of a mouth, and destitute of pseudopodia. Ex., Gregarina.

CLASS II. RHIZOPODA.—Simple or compound; destitute of a mouth; capable of putting forth pseudopodia.

CLASS III. INFUSORIA.—Generally with a mouth; no pseudopodia; with vibratile cilia or contractile filaments.

SUB-KINGDOM II.—CŒLENTERATA.

CLASS I. HYDROZOA.—Walls of the digestive sac not separated from those of the body cavity; reproductive organs external.

Sub-class 1. *Hydroida.*—Ex., Hydra. Tubularia (pipe-coralline). Sertularia (sea-fir).

Sub-class 2. *Siphonophora.*—Ex., Diphyes. Physalia (Portuguese man-of-war).

Sub-class 3. *Discophora.*—Ex., Naked-eyed Medusæ, or Jelly-fish.

Sub-class 4. *Lucernarida.*—Ex., Sea-nettles, or "Hidden-eyed" Medusæ.

CLASS II. ACTINOZOA.—Digestive sac distinct from the general cavity, but opening into it; reproductive organs internal.

Order 1. *Zoantharia.*—Ex., Sea-Anemones (Actinia). Reef-building corals.

Order 2. *Alcyonaria.*—Ex., Sea-pen. Red coral.

Order 3. *Ctenophora.*—Ex., Cestum (Venus's girdle).

SUB-KINGDOM III.—ANNULOIDA.

CLASS I. ECHINODERMATA.—Integument calcareous or leathery; adult radiate.
Order 1. *Crinoidea.*—Ex., Comatula.
Order 2. *Blastoidea.*—(Extinct.)
Order 3. *Cystoidea.*—(Extinct.)
Order 4. *Ophiuroidea.*—Ex., Brittle-star.
Order 5. *Asteroidea.*—Ex., Star-fish.
Order 6. *Echinoidea.*—Ex., Sea-urchins.
Order 7. *Holothuroidea.*—Ex., Sea-cucumbers.

CLASS II. SCOLECIDA.—Soft-bodied, cylindrical, or flat; nervous system not radiate; of one or two ganglia.
Order 1. *Tæniada.*—Ex., Tapeworms.
Order 2. *Trematoda.*—Ex., Flukes.
Order 3. *Turbellaria*—Ex., Planarians.
Order 4. *Acanthocephala.*—Ex., Echinorynchus.
Order 5. *Gordiacea.*—Ex., Hairworms.
Order 6. *Nematoda.*—Ex., Roundworms.
Order 7. *Rotifera.*—Ex., Wheel animalcules.

SUB-KINGDOM IV.—ANNULOSA.

DIVISION A. ANARTHROPODA.—Locomotive appendages not distinctly jointed or articulated to the body.
CLASS I. GEPHYREA.—Ex., Spoon-worms.
CLASS II. ANNELIDA. — Ex., Leeches (Hirundinidæ). Earth-worms (Oligochæta). Tube-worms (Tubicola). Sand-worms and Sea-worms (Errantia).
CLASS III. CHÆTOGNATHA.—Ex., Sagitta.
DIVISION B. ARTHROPODA. — Locomotive appendages jointed to the body.
CLASS I. CRUSTACEA.—Ex., Decapoda. Isopoda. Xiphosura. Cirripedia.
CLASS II. ARACHNIDA.—Ex., Podosomata (sea-spiders). Acarina (mites). Pedipalpi (scorpions). Arancida (spiders).

Class III. Myriapoda.—Ex., Centipedes.

Class IV. Insecta.— Ex., Anoplura (lice). Mallophaga (bird lice). Thysanura (spring-tails). Hemiptera. Orthoptera. Neuroptera. Diptera. Lepidoptera. Hymenoptera. Coleoptera.

SUB-KINGDOM V.—MOLLUSCA.

Division A. Molluscoida.—A single ganglion, or pair of ganglia; heart imperfect, or none.

Class I. Polyzoa.—Ex., Sea-mats (Flustra).

Class II. Tunicata.—Ex., Ascidia (Sea-squirts).

Class III. Brachiopoda.—Ex., Terebratula.

Division B. Mollusca Proper.—Three pairs of ganglia; heart of at least two chambers.

Class I. Lamellibranchiata.—Ex., Oyster. Mussel.

Class II. Gasteropoda. — Ex., Buccinium. Helix. Doris.

Class III. Pteropoda.—Ex., Cleodora.

Class IV. Cephalopoda.

Order 1. *Dibranchiata.*—Ex., Poulp. Paper Nautilus.

Order 2. *Tetrabranchiata.*—Ex., Pearly Nautilus.

CHAPTER XII.

THE MICROSCOPE IN ANIMAL HISTOLOGY.

In Chapter IX we described the elementary living substance, or bioplasm, from which all organized structures proceed, with an outline of its morphology, chemistry, and physiology. In Chapter X we treated of Vegetable Histology, or the elementary tissues and organs which pertain to vegetable life. We now consider the structure of formed material in animals, with special reference to the minute anatomy of the human body. Following the generalization of Dr. Beale, page 118, we may classify histological structures as follows :

A. INORGANIC AND ORGANIC ELEMENTS OR PABULUM.
    Resulting in
B. BIOPLASM; or, O. H. C. and N., with other chemical elements, *plus*, The cause of life.
    From this results:
C. FORMED MATERIAL, consisting of,
  I. CHEMICAL PRODUCTS; Organic Compounds, etc.
  II. MORPHOLOGICAL PRODUCTS. 1. Granules; 2. Globules; 3. Fibres; 4. Membrane.
    Forming *Tissues*. 1. Simple; 2. Compound.
    Arranged in *Organs*. 1. Vegetative; 2. Animal.

I. THE CHEMICAL PRODUCTS of Bioplasm are very numerous, and belong to the science of *Histo-Chemistry*. Our plan allows us to do little more than to enumerate the principal groups. It has already been stated that the true chemical structure of bioplasm, or living sarcode, (protoplasm in a living state) is unknown, since it is only possible to analyze the dead cell substance. Of the relation of the oxygen, hydrogen, carbon, and nitrogen, etc., which constitute its "physical basis," we can only speculate, or imagine. See *Chemistry of Cells and their Products*, page 122.

The chemical transformations of cell-substance into "formed material" consist chiefly, with water and mineral matter, of certain groups of *organic principles*, sometimes called albuminous or "protein" substances, and their nearer derivatives, as glutin-yielding and elastic matter, with fat and pigments. These materials are subject to constant secondary changes or transformations, since they are not laid down in the living body once for all. They are also subject to constant decay, or ultimate decomposition. *Histo-Chemistry* must, therefore, be always a difficult study, since we can rarely isolate the tissues for examination, nor always tell when a substance is superfluous aliment, formative or retrogressive material. From a limited number of formative or histogenic materials, we have a host of changed or decomposition products.

Frey's *Histology and Histo-Chemistry*, Stricker's *Hand-Book of Histology*, and Beale's *Bioplasm*, are among the most useful books to the student in this department.

Frey subdivides the groups of organic principles as follows:

I. *Albuminous or Protein Compounds.*—Albumen. Fibrin. Myosin. Casein. Globulin. Peptones. Ferments?

II. *Hæmoglobulin.*

III. *Formative (Histogenic) Derivatives from Albuminous Substances.*—Keratin. Mucin. Colloid. Glutin-yielding substances. Collagin and Glutin. Chondrigen and Chondrin. Elastin.

IV. *Fatty Acids and Fats.*—Glycerin. Formic acid. Acetic acid. Butyric acid. Capronic acid. Palmitic acid. Stearic acid. Oleic acid. Cerebrin. Cholesterin.

V. *Carbohydrates.*—The Grape-sugar group, Cane-sugar group, and Cellulose group; or Glycogen. Dextrin. Grape-sugar. Muscle-sugar. Sugar of milk.

VI. *Non-Nitrogenous Acids.*—Lactic. Oxalic. Succinic. Carbolic. Taurylic.

VII. *Nitrogenous Acids.*—Inosinic. Uric. Hippuric. Glycocholic. Taurocholic.

VIII. *Amides, Amido Acids, and Organic Bases.*—Urea. Guanin. Xanthin. Allantoin. Kreatin. Leucin. Tyrosin. Glycin. Cholin (Neurin). Taurin. Cystin.

IX. *Animal Coloring Matters.*—Hæmatin. Hæmin. Hæmatoidin. Urohæmatin. Melalin. Biliary pigments.

X. *Cyanogen Compounds.*—Sulpho-cyanogen.

XI. *Mineral Constituents.*—Oxygen, Nitrogen, Carbonic acid. Water. Hydrochloric acid. Silicic acid. Calcium compounds (Phosphate, Carbonate, Chloride, and Fluoride). Magnesium compounds (Phosphate. Carbonate. Chloride). Sodium compounds (Chloride. Carbonate. Phosphate. Sulphate). Potassium compounds (Chloride. Carbonate. Phosphate. Sulphate). Salts of Ammonium (Chloride. Carbonate). Iron and its Salts (Protochloride. Phosphate). Manganese. Copper.

The subject of *Histology* relates properly to cell-structure (already described, Chapter IX), and its morphological products, yet its close connection with Histo-chemistry renders the foregoing list of substances valuable to the student.

II. HISTOLOGICAL STRUCTURE is due to the formative power of bioplasm, or living cell-substance, and is not mere selection and separation from pabulum, or aliment, since from the same pabulum, and, so far as we can see, under the same circumstances, result tissues having different physical and chemical properties.

In our classification we have arranged the microscopic, or histological, elements of the tissues as Granules, Globules, Fibre, and Membrane.

Granules are minute particles of formed material.

Globules are small, homogeneous, round, or oval bodies. If composed of albuminous matter they are rendered transparent by acetic acid, and are dissolved by potash and soda. If consisting of fat they are soluble in ether and unaltered by acetic acid. If they are earthy matters they are dissolved by acids and unchanged by alkalies.

Fibres appear as fine lines, cylindrical threads, or flattened bands, parallel, or at various angles.

Membrane is an expansion of material. It may be transparent and homogeneous, and may be recognized by plaits or folds, which sometimes simulate fibres, or it may be granular, or bear earthy particles.

From these *elements* result the simple and compound *tissues*.

The Simple Tissues may be divided into

1. Cells with intermediate fluid, as Blood, Lymph, Chyle, Mucus, and Pus.
2. Epithelium and its appendages.
3. Connective Substances.—Cartilage. Fat. Connective tissue. Bone. Dentine.

The Compound Tissues are Muscle, Nerve, Gland, and Vascular tissues.

These are formed into *Organs*.

1. Vegetative.—The Circulatory, Respiratory, Digestive, Urinary, and Generative organs.
2. Animal.—The Bony, Muscular, Nervous, and Sensory apparatus.

We shall attempt a brief description of these tissues and organs, as illustrated by the microscope and modern methods of research.

I. SIMPLE TISSUES.

1. Cells with Intermediate Fluid.

I. *The Blood.*

The microscope shows blood to consist, especially in man and the higher animals, of red corpuscles, colorless corpuscles, and the fluid in which they are suspended.

1. *Blood Plasma, or Liquor Sanguinis.*—This is a colorless and apparently structureless fluid, but when removed from the body, fibrin separates from it in solid form. In small quantities of blood this is seen in delicate fibres crossing each other at various angles.

2. *Red-blood Corpuscles.*—These were first discovered by Swammerdam, in 1658, in frog's blood, and in that of man by Lewenhoek, in 1673. Malpighi is said to have first seen the actual circulation of blood in the web of a frog's

foot. The circulation may be readily observed by etherizing a frog, and expanding its foot by means of pins or thread, upon the stage of the microscope (Plate XVIII, Fig. 137). The circulation may also be seen in the lung, mesentery, or extended tongue, of the frog.

The red corpuscles of blood are flattened disks, which are circular in Mammals, except the camel and lama, which have elliptic disks. In birds, amphibia, and most fishes, the disks are elliptic. In a few fishes (the cyclostomata) they are circular. Their color depends on hæmoglobulin, which plays an important part in the exchange of respiratory gases. In man the disks are usually double-concave, with rounded edges. Out of the body they have a tendency to adhere, or run together, in chains, like rolls of coin (Plate XVIII, Fig. 138). In the elliptic disks of birds, etc , there is a distinct nucleus. The size of the disks varies. In man they are from 0.0045 to 0.0097 millimetre. The smallest disks are in the *Moschus Javanicus*, and the largest in *Siren lacertina*. In the latter they are from $\frac{1}{16}$ to $\frac{1}{30}$ millimetre.

It is estimated that in a cubic millimetre (about $\frac{1}{25}$th of an inch) of human blood there are 5,000,000 red corpuscles, having a surface of 643 millimetres.

After a variable time from their removal from the vessels they suffer contraction, and assume a stellate, or mulberry form (Plate XVIII, Fig. 139). This occurs more rapidly in feverish states of the system. On the warm stage they suffer still greater alterations. Indentations appear, which cause bead-like projections, some of which become fragments, having molecular motion (Plate XVIII, Fig. 139). The substance of red corpuscles is elastic and extensible, and may be seen in the vessels to elongate and curve so as to adapt themselves to the calibre of the vessels.

Electric discharges through the red corpuscles produce various changes of form. Alkalies dissolve, and acids cause a precipitate in them. They are tinged by neutral solutions of carminate of ammonia. One-half to 1 per cent. of salt added to the staining fluid causes the nuclei only of Amphibian corpuscles to be stained. Chloroform, tannin, and other reagents, produce various changes, which suggest a wide field of research connected with Therapeutics.

The old opinion of the structure of red corpuscles represented them as vesicles consisting of a membrane and its contents, but Max Schultze, in 1861, showed that a membrane was not constant. This may be verified by breaking them under pressure.

Brücke's experiment on the astringent action of boracic acid on the blood of *Triton*, repeated by Stricker and Lankester, shows the red corpuscles to possess a double structure. There is a body, called Œcoid; a porous, non-contractile, soft, transparent mass; and a retractile substance, or Zooid, containing the hæmoglobulin, which fills the interspaces of the Œcoid. The Zooid seems identical with simple cell-substance, or bioplasm.

3. *Colorless, or White Corpuscles.*—These appear to be simply masses of bioplasm of various sizes. Some are quite small, and many are larger than the red corpuscles. Their number is much smaller than the red disks, being about 1 to 350, or even less. In leucæmia and other diseases their relative number is much greater. In the blood of cold-blooded animals, and in that of vertebrata, if the normal temperature is continued by means of a warm stage, the amœboid motions are quite perceptible with a high magnifying power (Plate XVIII, Fig. 139). They may also be seen to take up small particles of matter into their interior, such as cinnabar, carmine, milk-globules, and even portions of the red globules.

Both red and white cells are forced through the uninjured walls of small vessels by impeded circulation, but the white cells thus migrate, by virtue of their vital con-

PLATE XVIII.

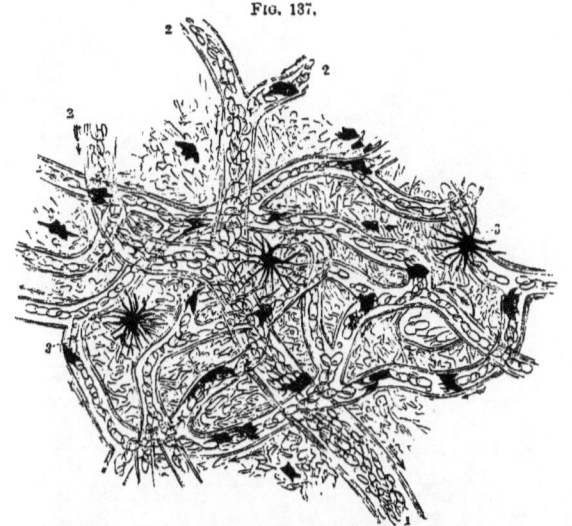

Fig. 137.

Capillary circulation in a portion of the web of a *Frog's* foot.

Fig. 138.

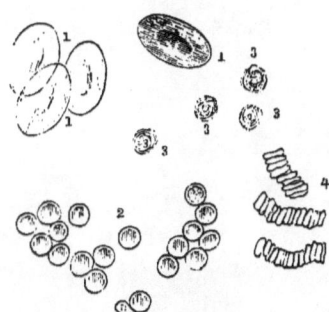

Blood-discs:—1, Elliptic Discs of Amphibia; 2, Human red-corpuscles; 3, White or lymph-corpuscle; 4, Rouleaux of red-discs.

Fig. 139.

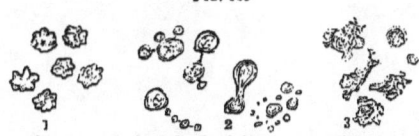

Alterations in form in blood-discs:—1, Stellate or mulberry form; 2, On warm stage; 3, Amœboid white-cell forms.

Fig. 140.

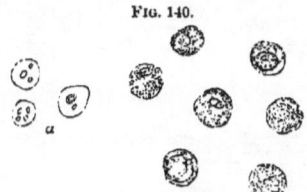

Pus-corpuscles:—*a*, with acetic acid.

Fig. 141.

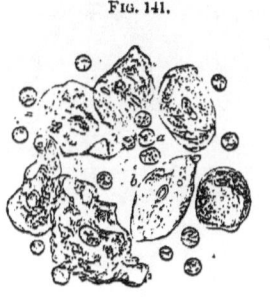

Mucous corpuscles and epithelium.

Fig. 142.

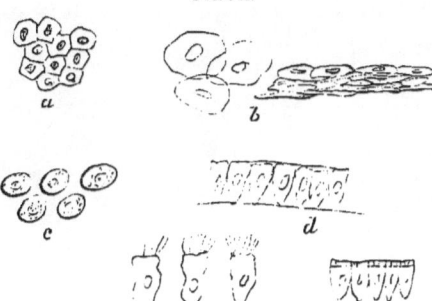

Varieties of Epithelium:—*a*, Tessalated; *b*, Squamous; *c*, Glandular; *d*, Columnar; *e*, Ciliated.

tractility, in the healthy body, and in greater numbers in diseased states; in some cases re-entering the lymphatic circulation, and in others penetrating into various tissues. The pus-corpuscles appearing in the vicinity of inflamed parts are shown by this discovery, made by Waller and Cohnheim, to be nothing but migratory lymphoid or white cells of the blood. The change of form and place of these amœboid cells is readily seen by placing a drop of frog's blood on a glass cover, and inverting it over a moist cell. As it coagulates, a zone of serum extends round the clot, in which the migrated cells will be found.

The colorless cells originate in the chyle and lymph-systems, although some may come from the spleen and the medulla of bones, multiplying in the blood itself, and they pass into red corpuscles. Transitional forms have been found in the general mass of blood, in the spleen, and in the marrow of bones.

The white or colorless cells of blood are identical with the cells of chyle, lymph, pus, mucus, and saliva. They are often described under the term *leucocytes* (white cells.)

The leucocytes of saliva (salivary corpuscles) and of pus contain granules or globules of formed material, which exhibit for some time a peculiar dancing movement (see page 120).

When at rest, or in a lifeless condition, the white cells are of spheroidal form, and generally exhibit granules and globules of fat. Acetic acid develops a nucleus, and sometimes splits it into several (Plate XVIII, Fig. 140).

II. *Lymph and Chyle.*

The vessels of the lymphatic or absorbent system receive the liquid part of the blood which has passed from the capillaries, together with the products of decomposition in the tissues, and return them to the circulation. The lymphatics of the intestinal canal receive during digestion a mixture of albuminous and fatty matters, which is known as *chyle*, and these vessels have obtained the name of *lacteals*. The cells in this fluid are leucocytes, identical with white cells in blood. They originate in the lymphatic glands and "Peyer's patches" of the intestine, and are the corpuscles of these organs which have been carried off by the fluid stream.

III. *Mucus.*

Is a tenacious semifluid substance which covers the surface of mucous membranes. It contains cast-off epithelial and gland-cells, and the *mucus corpuscle*, which, as we have before said, is identical with other leucocytes. *Synovial fluid* is of similar nature. It is now regarded as a transformation product of the epithelial cells, and not to originate as a secretion from special glands (Plate XVIII, Fig. 141).

2. Epithelium and its Appendages.

Epithelium (from $\epsilon\pi\iota$, upon, and $\theta\alpha\lambda\lambda\omega$, to sprout) is so called since it was formerly supposed to sprout from membrane. It is a tissue formed of cells more or less closely associated, which is found in layers upon external and internal surfaces. The cells are generally transparent, with vesicular, homogeneous, or granular nuclei, the latter being the remains of the original leucocyte or bioplast. In the older cells the nucleus is absent, the entire mass having been transformed.

The forms of epithelial cells vary according to situation or function. The original form is spheroidal, but changes by compression, etc.

1. Tessellated or pavement epithelium (Plate XVIII, *a*, Fig. 142). These are cells whose formed material is flattened, and which are united at their edges. They are sometimes hexagonal, and often polyhedral, in form.

Examples: Serous and synovial membranes; the posterior layer of the cornea; the peritoneal surface; the interior of bloodvessels, and shut sacs generally.

2. Squamous or scaly epithelium. The cells are flat, and overlap each other at the edges (Plate XVIII, *b*, Fig. 142).

Examples: Epidermis; many parts of mucous membranes, as the mouth, fundus of bladder, vagina, etc.

3. Glandular epithelium (Plate XVIII, *c*, Fig. 142). The cells are round or oval bioplasts, often polyhedral from pressure, and the formed material is often soft.

Examples: Liver cells, convoluted tubes of kidney, and interior of glands generally.

4. Columnar epithelium (Plate XVIII, *d*, Fig. 142). Cells cylindrical or oblong, arranged side by side. A bird's-eye view shows them similar to the tessellated form, hence they should be seen from the side.

Examples: Villi and follicles of intestine, ducts of glands, urethra, etc.

Some of the columnar or cylinder-cells have a thickened border or lid perforated with minute pores (Plate XVIII, *f*, Fig. 142). They are found in the small intestine, gall-bladder, and biliary ducts.

5. Ciliated epithelium (Plate XVIII, *e*, Fig. 142). These are cylindrical cells having vibratile cilia, whose motions produce a current in the surrounding fluid.

Examples: The upper and back nasal passages, the pharynx, bronchi, Fallopian tubes, etc.

The Hair.—Hairs are filiform appendages, composed of a modified epithelial tissue of rather complex structure. They originate in a follicle, which is a folding in of the skin. The shaft of the hair is the portion projecting above the skin, and the root is concealed in the hair-follicle. The bulb of the root is the rounded terminal part, which is hollow below, and rests on a papilla which rises from the floor of the follicle (Plate XIX, Fig. 143). Between the follicle and hair is a sheath, which is divided into an external and internal portion. The cells of the hair may be isolated by sulphuric acid or solution of soda. They overlap each other like tiles, so as to present undulating or jagged lines across the surface of a fresh hair. The felting property of wool depends on the looseness of this overlapping. Air-bubbles are often found in hair, especially in the medullary or axial portion, and give a silvery appearance to white hair. The granules of pigment are generally found in the cortical portion.

Nails are nothing more than modified cuticle, dependent for their growth on the vessels of the matrix or bed of the nail. Their epithelial cells may be demonstrated by soaking in caustic soda or potash.

Corns, warts, and *horn* have similar origin.

Enamel of the Teeth.—The minute structure of dental tissue will be described hereafter, but as the enamel is generally considered to be of epithelial origin, some account of it belongs here.

The edge of the jaw is first marked by a slight groove, known as the dental groove, and is covered with a thick ridge of epithelium, called the dental ridge (Plate XIX, Fig. 144, 1 *a*, 2 *a*). The epithelium grows down in a process which has been called the enamel germ (1 *d*). This becomes doubled by the upward growth of the dental germ (2, 3, *f*), which originates from connective tissue. The epithelial cells become transformed into enamel columns or prisms.

3. Connective Substances or Tissues.

The term connective tissue has been given to a variety of structures which probably start from the same rudiments, and have a near connection with each other. It is unfortunate that a name descriptive of function should be applied to structure, yet the present state of histology requires an account of substances thus called.

Connective tissues are all those which may be regarded

PLATE XIX.

Fig. 143.

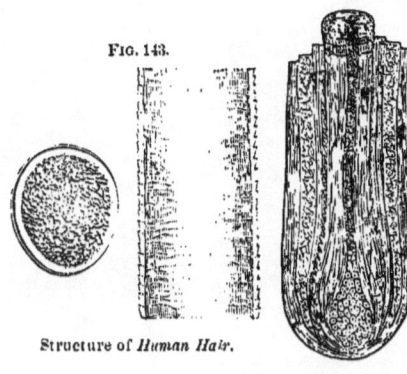

Structure of *Human Hair*.

Fig. 144.

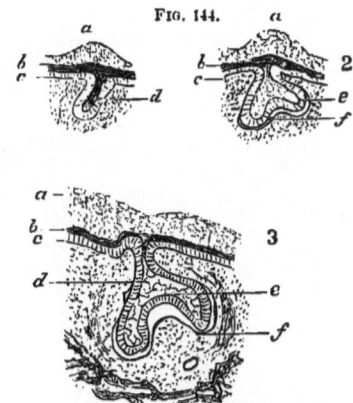

Development of the enamel:—*a*, dental ridge; *b*, young layer of epithelium; *c*, deep layer; *d*, enamel germ; *e*, enamel organ; *f*, dental germ.

Fig. 145.

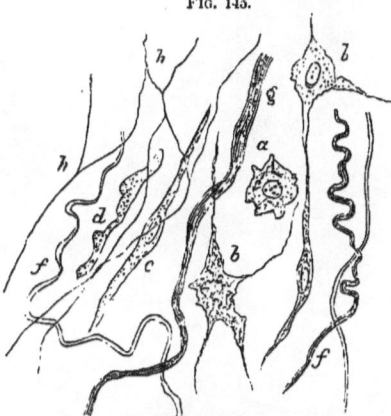

Connective-tissue elements. From the Frog's Thigh:— *a*, contracted cell; *b*, ramified; *c, d*, motionless granular cells; *f*, fibrillæ; *g*, connective-tissue bundle; *h*, elastic fibre net-work.

Fig. 146.

White Fibrous Tissue, from Ligament.

Fig. 147.

Yellow Fibrous Tissue, from Ligamentum Nuchæ of Calf.

Fig. 148.

Fatty Tissue.

as basement-membranes, supporting layers or investments for epithelial structures, blood, lymph, muscle, and nerves. It includes ordinary connective tissue (white and yellow fibrous tissues), cartilage, bone, corneal tissue, dentine, and fatty tissue.

Most of the difficulty found in the consideration of these tissues arises from discussions relative to the intercellular substance. Max Schultze and Beale agree in regarding it to originate from the protoplasm or bioplasm of cells.

The cells are, according to Frey, originally spheroidal, with vesicular nuclei, and between them is an albuminous intercellular substance—a product of the cells, or transformed cells—which usually undergoes fibrillation, while the cells become stunted, or develop into spindle-shaped or stellate elements. Calcification of the intercellular substance occurs in some of these tissues, as bone and dentine.

The cells of connective tissue present many varieties. Recklinghausen first observed migrating lymphoid cells or bioplasts in the cornea of the eye, the tail of the tadpole, the peritoneum, and in various other places. The exit of white corpuscles from the vascular walls renders it probable that these amœboid cells originate in the blood. Granular cells, of various forms—rounded, fusiform, and stellate—are also observed. Some of the stellate cells give off anastomosing branches. Pigment cells, filled with granular pigment, are also met with (Plate XIX, Fig. 145).

In its earliest stages, connective tissue consists of closely-compressed cells, but in the adult two principal forms have been distinguished; first, those networks and trabeculæ, developed from cells, which do not yield gelatin on boiling, and, secondly, fibrillar connective tissue composed of a gelatin-yielding substance. Of the first kind we notice the following varieties:

1. Independent masses of gelatinous or mucous tissue,

consisting of nucleated cells, giving off smooth anastomosing trabeculæ, as in the early stage of the vitreous humor of the eye and of the gelatinous tissue of the umbilical cord, etc.

2. Very delicate reticular tissue found in the eye and in the interior of nerve-centres.

3. A network filled with lymphoid cells (adenoid or cytogenous tissue) in the glands of the lymphatic system, and around the fasciculi of fibrillar connective tissue.

4. A coarser network in the ligamentum pectinatum of the human eye.

5. A tissue formed of fusiform and stellate cells, as in the interior of the kidneys

The second form referred to, or the *fibrillar connective tissue*, was the only form to which the term connective tissue was formerly applied. It is composed of gelatin-yielding fibrillæ, which may be split into skein-like portions of various breadth. (Plate XIX, Fig. 146.) Permanganate of potash stains it brown. Acetic and dilute mineral acids cause the tissue to swell so that the appearance of fibrillation is lost through compression, and the cells, or nuclei, are made manifest. Chloride of gold staining exhibits both fibrillæ and cells.

Elastic fibres (yellow elastic) (Plate XIX, Fig. 147) are apparent in all forms of connective tissue which have been made transparent by boiling, or acetic acid. They are non-gelatinizing, cylindric, slightly branched, or forming plexuses. In some fasciculi of fibrillar connective tissue, as seen after the action of acetic acid, elastic fibres appear in hoops, or spirals, around them. In the ligamentum nucleæ of the giraffe the elastic fibres are marked by transverse striæ, or cracks. Elastic fibres often form flattened trabeculæ, or are fused into elastic plates, or membranes, with foraminæ, as in arterial tunics.

The ligaments of the skeleton, the periosteum, perichondrium, aponeuroses, fasciæ, tendons, and generally all the tunics of the body, afford examples of the fibrillar connective tissue.

Fatty Tissue.—The loose connective tissue contains in various parts great numbers of cells filled with fat. Their form is round, or oval, and are often divided into groups, or lobules, by trabeculæ. (Plate XIX, Fig. 148.) Each lobule has its own system of bloodvessels, which divide into such numerous capillaries that the smaller groups, and even individual fat-cells, are surrounded by vascular loops. Sometimes the contents of the cells appear in needle-shaped crystals, often collected in a brush-like form. Fat-cells seem to be chiefly receptacles for the deposit of superabundant oleaginous nutriment, and are analogous to the starch-cells in vegetables.

Cartilage.—This is formed of cells in an originally homogeneous intercellular substance. The only difference between what was formerly distinguished as cartilage and fibro-cartilage is that the matrix or intercellular substance of the latter is fibrous.

The cells, or cartilage-corpuscles, are nucleated, and lie in cavities of various sizes and form in the matrix (Plate XX, Fig. 149). Two nuclei often appear in one cell. It is yet a question whether the capsule and matrix are the secretion of the cells which has become solid, or a part of the body of the cell which has undergone metamorphosis.

The multiplication of cartilage-cells is endogenous. By segmentation, two, four, or a whole generation of daughter-cells, so called, may lie in the interior of a capsule. In this way growing cartilage may acquire a great number of elements.

In the ear of the mouse, etc., we observe a form of cartilage which is wholly cellular, and possesses no matrix (Plate XX, Fig. 150).

Bone, or osseous tissue, is formed secondarily from metamorphosed descendants of cartilage or connective-tissue cells, and is the most complex structure of this group. It consists essentially of stellate ramifying spaces containing cells, and a hard, solid, intermediate substance. The latter is composed of glutinous material rendered hard by a mixture of inorganic salts, chiefly of calcium.

As all bones are moulded first in cartilage it was natural to conceive that they were developed by a transformation of cartilage. Much variety of opinion still exists respecting the process, but it is generally conceded that although cartilage may undergo calcification, true bone is not formed until the cartilage is dissolved. New generations of stellate cells appear in a matrix, which is first soft and then calcified. New bone may also grow from the periosteum by means of a stratum of cells called osteoblasts. The details of the process are too extensive for a treatise like the present. If sections of growing bone are decalcified with chromic acid and treated with carmine, the osteoblastic layers and adjacent youngest bony layer acquire an intensely red color, while the rest of the tissue, except the bone-corpuscles, remains uncolored.

Fine sections cut from a long bone longitudinally and transversely will show the microscopic structure, consisting of the *Haversian canals* (Plate XX, Fig. 151, *a*) surrounded with *concentric lamellæ* of compact structure (*b, b*). There are also intermediate and periosteal lamellæ (*c, d*). The cavities containing the bone-cells, or bioplasts (*e, e,*) are of various sizes, from 0.0181 to 0.0514 millimetres long, and from these *lacunæ* run the *canaliculi* in an irregular radiating course (*f, f*). In a balsam-mounted specimen these hollows sometimes retain air, by which the structure is rendered more apparent.

Dentine is the structure of which the teeth are most largely composed. It consists of minute tubes filled with bioplasm, which radiate from the central cavity of the tooth, the interspaces between the tubes being solidified by earthy salts so that the tissue is harder than bone.

Histologically a tooth may be said to be made of three

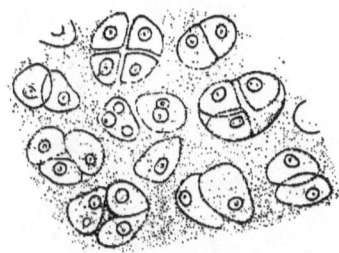

Fig. 149.

Section of the Branchial *Cartilage* of Tadpole.

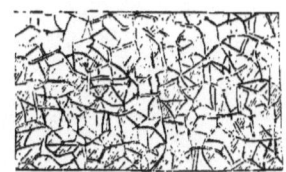

Fig. 150.

Cellular Cartilage of Mouse's Ear.

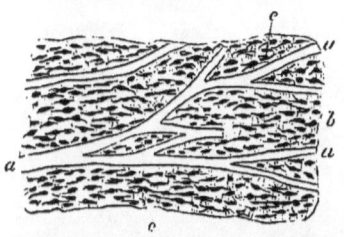

Fig. 151.

Longitudinal and transverse section of *Bone*:—a, Haversian canals; b, concentric lamellæ; c, intermediate; d, periostial lamellæ; e, bone-cells; f, canaliculi.

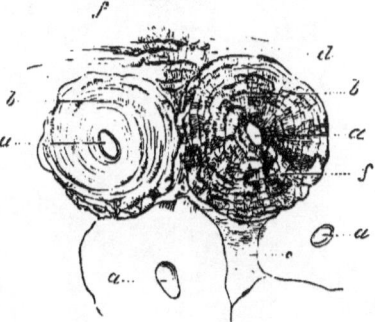

Fig. 152.

Vertical section of *Human Molar Tooth*:—1, enamel; 2, cementum or crusta petrosa; 3, dentine, or ivory; 4, osseous excrescence, arising from hypertrophy of cementum; 5, pulp-cavity; 6, osseous lacunæ at outer part of dentine.

Fig. 153.

Involuntary Muscular-fibre.

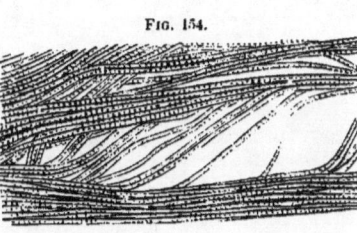

Fig. 154.

Striated Muscular-fibre, separated into fibrillæ

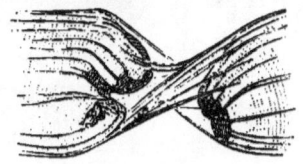

Sarcolemma.

kinds of tissue: the *cement*, a bony substance, coating the root of the tooth, containing bone-cells and canaliculi, but no Haversian canals, the pulp in the central cavity of the tooth serving for the nutrition of the organ, as a large Haversian canal; the *dentine*, or ivory, constructed as above described; and the *enamel*, covering the crown, and consisting of columns or prisms, often hexagonal, which are the hardest and densest structures of the body (Plate XX, Fig. 152).

The development of enamel from epithelium has been referred to on page 192. The dental germ corresponds to a papilla of the mucous membrane, and in an early stage is covered by delicate stratified cells—the dentine cells, or odontoblasts—which produce dentine. Teeth are thus produced abnormally in other situations besides the jaws, as in ovarian cysts, etc.

Before the development of the first, or milk teeth, the rudiments of the permanent teeth exist as a fold or leaf of epithelium springing from the enamel germ.

II. COMPOUND TISSUES.

1. *Muscle.*—This is the tissue by which the principal movements of the body are performed. It consists of fibrin, which is endowed with special contractile power. It is of two kinds, the voluntary, pertaining to organs of voluntary motion, and the involuntary, found in situations which are not under the control of volition, as the coats of bloodvessels, alimentary canal, uterus, and bladder. The fibres of voluntary muscles are marked with transverse striæ. Involuntary muscular fibres are smooth, except in a few instances, as the fibres of the heart and some of those in the œsophagus, which are striated.

The fibres are connected with and invested by connective tissue, and arranged in parallel sets, with vessels and nerves in the intervals, and are attached to the parts they are designed to move by tendon, aponeuroses, or some form of fibrous tissue. The organs or muscles thus formed are generally solid and elongated, but sometimes expanded.

Involuntary or *unstriped muscular fibres* are flat bands or spindle-shaped fibres with nuclei, which may be regarded as the remains of the formative bioplasm (Plate XX, Fig. 153). They are usually transverse, or interlace with each other on the walls of cavities and vessels. In the heart the fibres, though involuntary, are striped and branching. *Striped fibre* varies from $\frac{1}{5 0}$th to $\frac{1}{1800}$th inch in diameter. It is largest in insects, in which individual fibrils may be readily obtained, especially from the thoracic muscles. They are generally found in bundles of fibrils, splitting longitudinally or in disks, and each bundle is inclosed in a sheath or sarcolemma (Plate XX, Fig. 154).

The transverse striation of muscle is subject to much variation, and the precise nature of the sarcous elements which produce the appearance is yet a matter of dispute, but in all probability the ultimate elements are sarcous prisms or particles imbedded in a homogeneous mass, and by their mutual attraction, excited by various stimuli, the contraction of the fibre takes place.

For the purpose of observation, the connective tissue may be removed from muscular fibre by gelatinizing it with dilute sulphuric acid, and dissolving it at a temperature of 104° F. The nuclei of muscular fibre are seen after treating with acetic acid, and may be stained with carmine fluid, etc.

2. *Nerve-tissue.*—The term nerve was applied by the ancients to tense cords, as bow-strings, musical strings, etc., and was appropriated to the fibres now called nerves, because they deemed them to operate by tremors, vibrations, or oscillations, another instance of wrong naming of structure from an opinion respecting function. Hippocrates, Galen, and others, however, thought nerves were

hollow tubes, conveying fine ethereal fluids, termed animal spirits.

Nervous matter is soft, unctuous, and easily disturbed, hence it is necessary to examine it while fresh. Histologically it is divided into fibres and cells, imbedded in connective tissue.

Nerve-fibres are of two kinds, the medullated, or dark-bordered threads, and the pale, or non-medullated. Medullated fibres consist of a delicate envelope of connective tissue, called the neurilemma or primitive sheath, an axis-cylinder or albuminous portion, extending down the centre, and a portion composed of a mixture of albumen, cerebral matter, and fat, surrounding the axis-cylinder (Plate XXI, Fig. 155, A, B, C). This latter is the medullary sheath, or white substance of Schwann. It changes rapidly, so as to coagulate and become granular. Alkalies render it fluid, so as to exude in fat-like drops. Absolute alcohol, chromate of potass and collodion, contract the sheath, so as to permit the axis-cylinder, which is the essential part of the nerve, to protrude (Plate XXI, E, Fig. 155). Anilin, carmine, nitrate of silver, and chloride of gold stain the axis, while osmic acid blackens only the medullary sheath.

Non-medullary or pale nerve-fibres are regarded as embryonic or developmental forms (Plate XXI, D, Fig. 155). The ganglionic fibres of the sympathetic (Remak's fibres) are flat, homogeneous bands, with round or oval nuclei. Some have considered them as formed of connective tissue, but their nervous character is generally conceded.

Schultze and others regard the axis-cylinder as made up of extremely delicate fibrillæ.

Nerve-cells, or ganglion corpuscles, are of two kinds, those without and those with processes. The first are called apolar, and the latter unipolar, bipolar, or multipolar, according to the number of ramifications. The cells are nucleated, and inside the nucleus is usually another, the nucleolus. Dr. Beale discovered certain ganglion-cells in the sympathetic of the tree-frog (in the auricular septum of the heart), one of whose poles is encircled spirally by the others (Plate XXI, Fig. 156).

The ultimate structure of ganglia or nervous knots, and the relation of the fibres to the cells, opens a wide field of research. In the muscle of the heart, etc., many of these ganglia seem to form special nervous systems. Dr. Beale has described the nerves ramifying on the capillaries and involuntary muscular fibrils of the terminal arteries as a self-regulating mechanism for the distribution of blood (Plate XXI, Fig. 157). Thus, if a tissue receives excess of pabulum, the capillary nerve-fibre is disturbed and transmits a change to the ganglion, and thence through the efferent nerve to the muscular fibres of the artery, and *vice versâ*.

Meissner has shown many ganglionic plexuses in the submucous coat of the alimentary canal. Another system of the same kind, called the *plexus myentericus*, was discovered by Auerbach between the muscular layers of the intestinal tube. Similar plexuses exist in other organs.

As to the peripheral termination of nerve-fibres, there is still considerable discussion. Most of the German histologists consider the nerves of voluntary muscles to terminate in end plates, in which the neurilemma becomes continuous with the sarcolemma of the muscular fibre. Dr. Beale maintains that there is a plexus of minute nerves over the fibrils. In some of my own preparations, especially some stained with soluble Prussian blue, a disk formed of a plexus of excessively minute nerve-fibres is observed, from which tortuous branches go to other muscle-fibres.

In the cornea, Cohnheim and Klein have traced fine nerve-fibres to the epithelial cells of the conjunctiva, by means of chloride of gold staining.

3. *Glandular tissue* consists of a fine transparent mem-

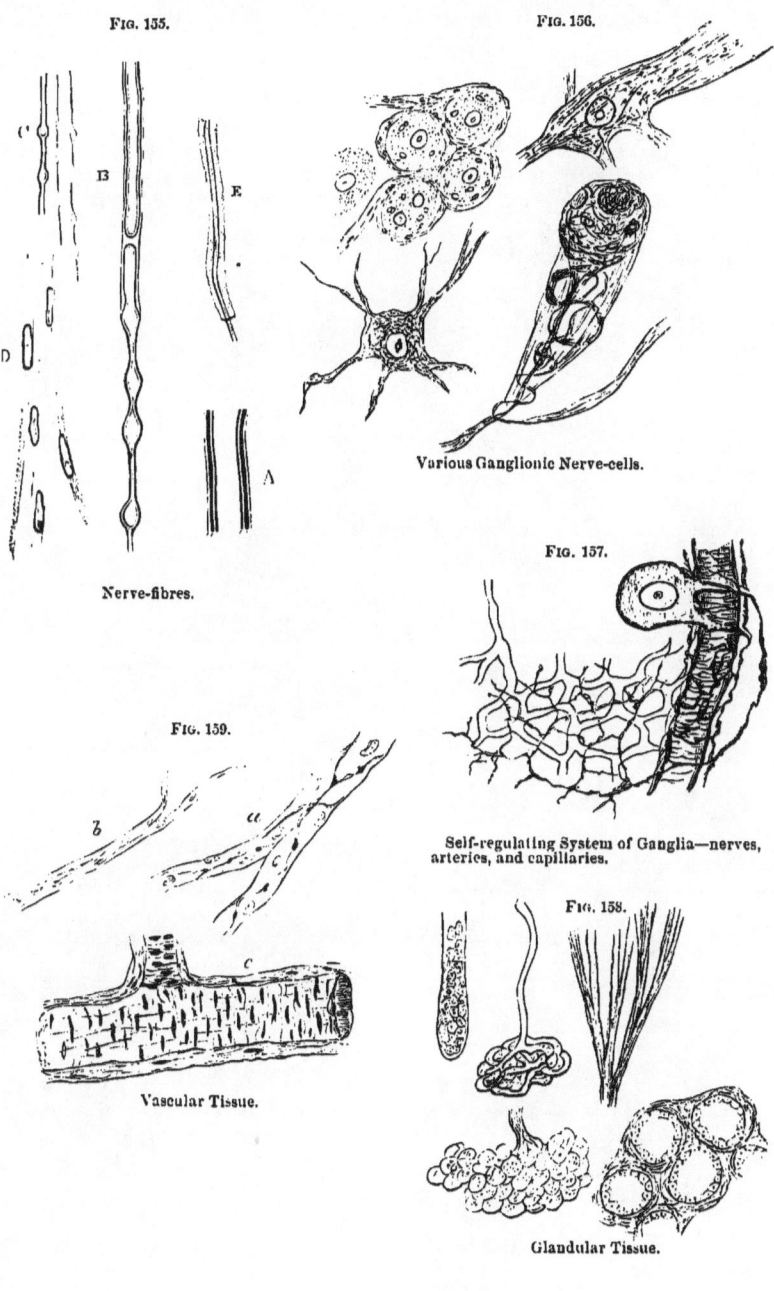

Fig. 155. Nerve-fibres.

Fig. 156. Various Ganglionic Nerve-cells.

Fig. 157. Self-regulating System of Ganglia—nerves, arteries, and capillaries.

Fig. 159. Vascular Tissue.

Fig. 158. Glandular Tissue.

brane, through which the plasma transudes, and cells of glandular epithelium. A vascular network exists on the surface of the membrane, from which the material of the secretion is obtained. This membrane may be a simple follicle, or tube, as in the mucous membrane, or system of tubes, as in the kidneys, a convoluted tube, a simple open vesicle, a racemose aggregation of vesicles, or a close capsule which discharges itself by bursting. (Plate XXI, Fig. 158).

4. *Vascular Tissue.*—The smallest bloodvessels and lymphatics, called capillaries, are minute tubes, consisting of a series of flattened epithelial cells, and containing stomata, or openings through which white or red blood-corpuscles may occasionally pass (Plate XX, Fig. 159, *a, b*). The larger trunks have, in addition to the cellular layer, one of longitudinally striated connective tissue, a middle coat containing transverse muscular fibres, and an external coat of connective tissue (Plate XXI, Fig. 159). The distribution of the capillary bloodvessels is various, according to the nature or function of the organ or tissue in which they are found.

Development of the Tissues.

It has been stated, page 125, that reproduction in the higher animals consists of an ovum fecundated by contact with a sperm-cell, or spermatozoid. The ovum consists of a *germinal vesicle*, containing one or more *germinal* spots, and included within a *vitellus* (a yelk) which is surrounded by a *vitelline membrane*, which may have additional investments in the form of layers of albumen and of an outer coriaceous or calcified shell.

The first step in the development of the embryo is the division of the vitelline substance into *cleavage-masses*, at first two, then four, then eight, etc. This process of yelk-division may affect the whole yelk or a part of it, and results in the formation of a *blastoderm*, or embryogenic tissue. This rudimentary embryonic tissue consists of three layers of cells, or germinal plates. The upper is the *corneous* layer, or epiblast, the middle one the *intermediate* plate, or mesoblast, and the lower the *intestinal glandular* layer, or hypoblast (Plate XXI, Fig. 160). From these the various tissues and organs are developed.

The outer plate produces the epithelium of the skin and its appendages, with the cellular elements of the glands of the skin, mammæ, and lachrymal organs. By a peculiar folding over the axis this plate also produces the elements of the brain and spinal cord, and the internal parts of the organs of special sense. The physiological significance of this layer is, therefore, very great.

The lower stratum of the blastoderm supplies the epithelium of the digestive tract, and the cellular constituents of its various glands, together with the liver, lungs, and pancreas.

The middle layer supplies the material for many structures. The whole group of connective substances, or tissues of support; muscular tissue; blood and lymph, with their containing vessels; lymph-glands, including the spleen, etc., all arise from this. The epithelial cells of such tubes and cavities as originate in this layer are regarded as different from those of true glands, and are more permeable to fluids. They have been termed false epithelium, or *endothelium*.

The following description, by Professor Huxley, will enable the student to form an idea of the general process of development. A linear depression, the *primitive groove*, makes its appearance on the surface of the blastoderm, and the substance of the mesoblast along each side of this groove grows up, carrying with it the superjacent epiblast. Thus are produced the two *dorsal laminæ*, the free edges of which arch over toward one another, and eventually unite, so as to convert the primitive groove into the cerebro-spinal canal. The portion of the epiblast which lines this, cut off from the rest, becomes thickened, and takes on the structure of the brain, or *encephalon*, in the region of the head; and of the spinal cord, or *myelon*, in the region of the spine. The rest of the epiblast is converted into the epidermis.

The part of the blastoderm which lies external to the dorsal laminæ forms the *ventral laminæ;* and these bend downward and inward, at a short distance on either side of the dorsal tube, to become the walls of a ventral or visceral tube. The ventral laminæ carry the epiblast on their outer surfaces, and the hypoblast on their inner surfaces, and thus, in most cases, tend to constrict off the central from the peripheral portions of the blastoderm. The latter, extending over the yelk, incloses it in a kind of bag. This bag is the first formed and the most constant of the temporary, or fœtal appendages of the young vertebrate, the *umbilical vesicle*.

While these changes are occurring, the mesoblast splits, throughout the regions of the thorax and abdomen, from its ventral margin, nearly up to the *notochord* (which has been developed, in the meanwhile, by histological differentiation of the axial indifferent tissue, immediately under the floor of the primitive groove) into two *lamellæ*. One of these, the *visceral lamella*, remains closely adherent to the hypoblast, forming with it the *splanchnopleure*, and eventually becomes the proper wall of the enteric canal; while the other, the *parietal lamella*, follows the epiblast, forming with it the *somatopleure*, which is converted into the parietes of the thorax and abdomen. The point of the middle line of the abdomen at which the somatopleures eventually unite, is the *umbilicus*.

The walls of the cavity formed by the splitting of the ventral laminæ acquire an epithelial lining, and become the great *pleuroperitoneal* serous membranes (Huxley's *Anatomy of Vertebrated Animals*).

In addition to the umbilical vesicle, above described as

a temporary appendage, the fœtus has other special structures, derived from the blastoderm. Thus the somatopleure grows up over the embryo and forms a sac filled with clear fluid, the *amnion*. The outer layer of the sac coalesces with the vitelline membrane to form the *chorion*. The *allantois* begins as an outgrowth from the mesoblast. It becomes a vesicle, and receives the ducts of the *primordial kidneys* or *Wolffian bodies*, and is supplied with blood from the two hypogastric arteries which spring from the aorta. The allantois is afterwards cast off by the contraction of its pedicle, but a part of its root is usually retained, and becomes the permanent urinary bladder. In the Mammalia the allantois conveys the embryonic vessels to the internal surface of the chorion, whence they draw supplies from the vascular lining of the uterus.

Foster and Balfour recommend that the study of embryonic development should commence with the egg of a fowl taken at different times from a brooding hen, or an artificial incubator. The egg should be placed on a hollow mould of lead in a basin, and covered with a warm solution of salt (7.5 per cent.). It should be opened with a blow, or by filing the shell. With the naked eye or simple lens, lying across the long axis of the egg, may be seen the *pellucid area*, in which the embryo appears as a white streak. The mottled *vascular area*, with the blood-vessels, and the *opaque area* spreading over the yelk, may be observed. The blastoderm may be cut out with a sharp pair of fine scissors, floated into a watch-glass, freed from vitelline membrane and yelk, and removed (under the salt solution) to a glass slide. A thin ring of putty may then be placed round the blastoderm, which is covered with salt solution, and the thin glass cover put on. With a low-power objective many of the details of structure may be seen in an embryo of thirty-six to forty-eight hours incubation, as the heart, the neural tube, the first cerebral vesicles, the folds of the somatopleure and splanchnopleure, the provertebræ, etc.

To prepare sections of the embryo, it must be first hardened by placing the slide containing it in a solution of 1 per cent. chromic acid for twenty-four hours. From this it should be removed to one of 3 per cent. for twenty-four hours more; then for a similar time in alcohol of 70 per cent., then in alcohol of 90 per cent., and lastly in absolute alcohol, where it may remain till required for section. Sometimes picric or osmic acid is used for hardening. The embryo may be stained by placing it in Beale's carmine fluid for twenty-four hours, and then replacing it in absolute alcohol for a day before it is cut. It may also be stained with hæmatoxylin if preferred. The specimen may be imbedded in paraffin, wax, and oil, or a mixture of four parts of spermaceti to one part of cocoa butter or castor oil. If there are cavities in the object, it is best to saturate it first with oil of bergamot. A little melted spermaceti mixture is poured on the bottom of a small paper box, and when solid the embryo is placed flat on it, the superfluous oil removed as far as possible, and the warm mixture poured on. Bubbles can be removed with a hot needle. A mark should be made of the exact position of the embryo. Sections may be cut with the section-cutter or a sharp razor, and if the spermaceti mixture is used, the razor should be moistened with olive oil. The sections should be floated from the razor to the slide, and treated with a mixture of four parts turpentine and one of creasote. They may then be mounted in balsam or dammar varnish.

The most instructive transverse sections of an early embryo will be through the optic vesicles, the hind brain, the middle of the heart, the point of divergence of the splanchnopleure folds, the dorsal region, and a point where the medullary canal is still open. For the unincubated blastoderm only one section, through the centre, is required to show the formative layers. In the later stages dissection is required, and is best performed with embryo preserved in spirit. If living embryos are placed in spirit, a natural injection of the vessels may be obtained.

III. ORGANS OF THE BODY.

Anatomists usually group the organs into systems, as the osseous, muscular, nervous, vascular systems, etc., but for histological study a classification based on physiological considerations may be more convenient for the student.

I. VEGETATIVE ORGANS.

1. *Nutritive*, or organs pertaining to the absorption and distribution of pabulum, including the digestive and circulatory organs.

The mucous membrane of the intestinal canal contains many follicles and glands, whose secretions serve important offices in the preparation of the food. These will be referred to in the next section. The epithelium of the intestinal canal is columnar, except in the œsophagus, where it is laminated. Beneath the glandular layer of the stomach is a stratum of fibrous connective tissue and muscle fibres in two layers, an internal with transverse, and an external with longitudinal fibres. The tissue of the small intestine beneath the epithelium is reticular connective, entangling lymphoid cells. The structure of the large intestine is similar to that of the stomach. The villi of the small intestine begins at the pylorus, flat and low at first, but becoming conical, and finally finger-like in shape. The epithelium of the villi are columnar, with a thickened and perforated edge (Plate XXII, Fig. 161). Between the epithelial cells of the villi, peculiar "goblet-cells" are often found, which Frey supposes to be decaying cells. The reticular connective tissue of each villus is traversed by a vascular network, a lymphatic canal or lacteal, and delicate longitudinal muscular fibres. If the villus is unusually broad, there may be more than one lacteal. The *lacteals* absorb the fluid known as chyle. They are blind ducts, and nitrate of silver injections show them to have the same structure as other lymphatics.

The *lymphatic radicles* are widely disseminated through all the tissues and organs of the body. They take up nutritive fluids, either from the alimentary canal, or such as have transuded from the capillaries into the interstices of the body, mingled with the products of decomposition, and convey them into the general circulation. Hyrtl's method of demonstrating these radicles is by passing a fine canula into the tissue containing lymphatics and forcing the injection by gentle pressure. They are either networks, analogous to capillaries, or blind passages which unite in reticulations. The structure of the vessels has already been described, page 201. Lymphatics and capillaries do not communicate directly. A lymph-canal may be surrounded by capillaries, or run alongside of a capillary, or a lymphatic sheath may envelop a bloodvessel. This latter plan is seen in the nervous centres, and has been called by His the *perivascular canal system*.

The larger lymphatic trunks are interrupted by nodular and very vascular organs, the *lymphatic glands*. These consist of the reticular connective tissue already described, surrounded by an envelope of ordinary fibrous tissue. One or more afferent lymphatic vessels penetrate the capsule, or envelope, and similar efferent vessels make their exit from the other side. Frey describes these glands as consisting of a cortical portion, follicles, and a medullary portion composed of the tubes and reticular prolongations of the follicles (Plate XXII, Fig. 162). There is a complicate system of communication between the follicles. The afferent vessel opens into the investing spaces of the follicle. These lead into the lymph-passages of the medullary portion, from the confluence of which the radicles of the efferent vessels are formed. The lingual follicular glands and tonsils, the solitary and agminated glands of the intestine (Peyer's patches), the thymus, and the spleen have a similar structure, and are called *lymphoid organs*.

In the *thoracic duct* the epithelium is inclosed in several layers of fibrous membrane. The latter contains transverse muscular fibres. The *heart*, although an involuntary muscular organ, has striated muscular fibres. These fibres are not, like other striped muscles, collected into bundles, but are reticular. The heart, like other organs, is supplied with lymphatics and bloodvessels. The cardiac plexus of nerves consists of branches from the vagus and sympathetic. Numerous microscopic nervous ganglia also occur, especially near the transverse groove and septum of the ventricles. It is thought that these are the chief centres of energy, so that the heart pulsates after its removal from the body. It has also been shown recently that the sympathetic and vagus filaments are in antagonism, so that stimulation of the vagus interrupts the motor influence of the sympathetic, and may bring the heart to a standstill in a condition of diastole.

The structure of bloodvessels has been described under the head of vascular tissue. No special boundary exists between capillaries and the arteries and veins. The arrangement of the capillaries, however, is various, and often so characteristic that a practiced eye can generally recognize an organ or tissue from its injected capillaries. (Plate XXII, Figs. 163 to 168.) For methods of injecting, see page 64. Capillaries form either longitudinal or rounded meshes. The muscular network, etc., is extended, while fat-cells, the alveoli of the lungs, lobules of liver, capillary loops of papillæ in skin and mucous membranes, outlets of follicles, etc., present a more or less circular interlacement. The capillary tube lies external to the elementary structure, and never penetrates its interior.

2. *Secretive Organs.*—True secretions serve important offices in the organism: as the materials of reproduction;

PLATE XXII.

Fig. 161.

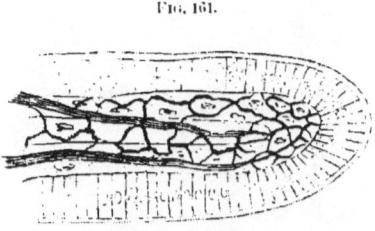

Intestinal Villus.

Fig. 162.

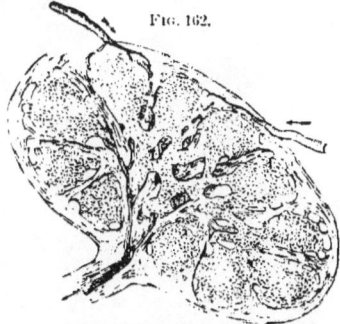

Lymphatic Gland.

Fig. 163.

Capillary net-work around *Fat-cells*.

Fig. 164.

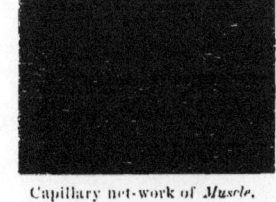

Capillary net-work of *Muscle*.

Fig. 165.

Distribution of Capillaries in *Mucous Membrane*.

Fig. 166.

Distribution of Capillary bloodvessels, in *Skin of Finger*.

Fig. 167.

Villi of Small Intestine of Monkey.

Fig. 168.

Arrangement of the Capillaries of the air-cells of the *Human Lung*.

milk from the mammary gland; saliva, gastric juice and pancreatic fluid for digestion; mucus, sebaceous matter, tears, etc. Excretions result from waste or decomposition, and are incapable of further use; as carbonic acid, separated by the lungs; urea, uric acid, etc., by the kidneys; saline matters, from kidneys and skin; lactic acid, portions of bile, and some of the components of fæces.

The *sweat glands* in the skin are simply convoluted tubes lined with glandular epithelium and surrounded by a basket-like plexus of capillaries. The *sebaceous glands* are racemose, and often open into the hair-follicles.

The *salivary glands* are complex mucous glands, and the saliva secreted by them is a complex mixture. The terminal nerves of the submaxillary gland have been traced to the nuclei of the gland-cells.

The lingual glands, and parotid, partake of the nature of lymphoid organs. The glands of the œsophagus are racemose. In the stomach there are two kinds, the *peptic*, and gastric *mucous* glands. The peptic glands are blind tubes closely crowded together over the mucous membrane, lined with columnar epithelium near their openings, and gland-cells below. The mucous glands are numerous near the pylorus, and are usually branching tubes. The capillaries are arranged in long meshes about the peptic glands, and form a delicate network in the submucous tissue. Numerous lymphatic radicles communicate with lymph-vessels below the peptic glands.

The *small intestine* contains the racemose *glands of Brunner* and the tubular *follicles of Lieberkuhn*, together with the lymphoid follicles known as the solitary and agminated *glands of Peyer*. The glands of Brunner are confined to the duodenum, and their excretory duct and gland vesicle are lined by columnar epithelium. Lieberkuhn's follicles are found in great numbers all over the small intestine. Peyer's patches are most numerous in the ileum. They are accumulations of solitary glands, and their structure is similar to the follicles of a lymphatic gland. The gland vesicles of the *pancreas* are roundish, and like other salivary glands it is invested with a vascular network with rounded meshes.

The liver is the largest gland connected with nutrition. Few animals are without a liver or its structural equivalent. In polyps the liver is represented by colored cells in the walls of the stomach cavity. In annelids the biliary cells cluster round cæcal prolongations of the digestive cavity. In crustacea the liver consists of follicles, and in insects of tubes, opening into the intestine. In all cases the essential elements are glandular cells containing coloring matter, oil, etc. In vertebrates some parts of the structure have not been decided upon without controversy.

In man the liver is a large, solid, reddish-brown gland, about twelve inches across, and six or seven inches from anterior to posterior edge, and weighing three or four pounds, situated in the right hypochondrium, and reaching over to the left. It is divisible into right and left lobes by the broad peritoneal ligament above, and the longitudinal fissure beneath. From the latter a groove passes transversely on the right side, lodging the biliary ducts, sinus of the portal vein, hepatic artery, lymphatics, and nerves, which are enveloped in areolar tissue, called the capsule of Glisson. From this groove ramifications of the portal canal extend through the liver, so numerous that no part of the hepatic substance is further than one-thirtieth of an inch from them. These ramifications carry the branches of the portal vein from which the capillary plexus surrounding the lobules begin, together with the bile-ducts, hepatic artery, etc.

The hepatic lobules are readily distinguished by the naked eye in many mammals, as the hog, but less easily in human liver. They consist essentially of innumerable gland-cells, and a complex network of vessels which tend towards the centre of the lobule, where their confluence

PLATE XXIII.

Fig. 169.

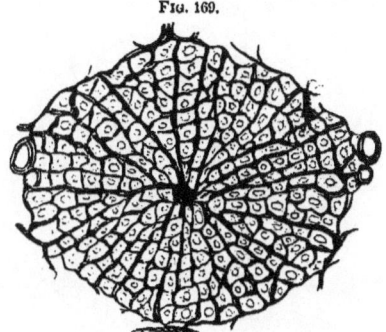

Lobule of Liver.

Fig. 170.

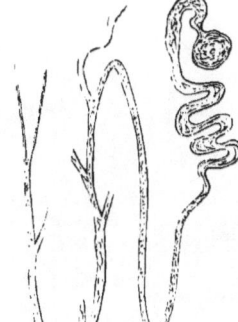

Uriniferous Tubes of Kidney.

Fig. 171.

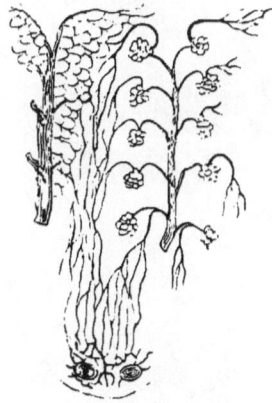

Blood-vessels of Kidney.

Fig. 173.

Tactile Papillæ.

Fig. 172.

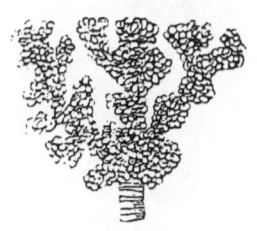

Alveoli of Lung.

Fig. 174.

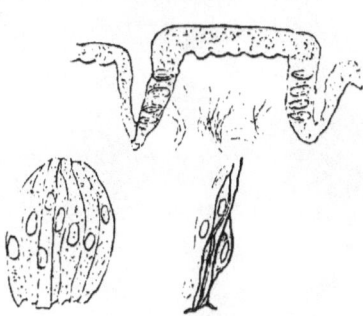

Taste-buds.

forms the radicle of the hepatic vein; while externally the lobules are bounded by branches of the portal vein and biliary canals (Plate XXIII, Fig. 169). The hepatic artery nourishes the proper connective tissue of the organ, and its venous radicles return the blood to the portal vein. The liver or bile-cells lie between the meshes of the capillaries, and are irregularly polyhedral from pressure, soft, granular, and nucleated. Brown pigment-granules and fatty globules are also found in the cells, and in disease in increased quantity. These bile-cells are inclosed in a delicate reticulated membrane, and Hering considers them to have a plexus of fine bile-ducts around them.

The kidneys are two large bean-shaped organs, each covered with a thin but strong fibrous envelope or tunic, which is continuous round the organ to the hilus, where the ureter leaves the gland and the bloodvessels enter. Even with the naked eye we may distinguish in a section of kidney the external granular cortex and the fibrous or striped medullary portion. The lines of the latter converge towards the hilus, and generally in a single conoid mass; but in man and some other animals this is divided into sections, called the pyramids, and between them the cortical substance is prolonged in the form of septæ, while both portions contain interstitial connective tissue. Both the cortical and medullary portions contain long branching glandular tubes, called the uriniferous tubes. In the medullary part these tubes are straight and divide at acute angles, while in the cortex they are greatly convoluted and terminate in blind dilatations, the capsules of Bowman. Staining with nitrate of silver shows the capsules to be lined with delicate pavement-epithelium. The convoluted tubes proceeding from the capsules, containing thick granular gland-cells, after numerous windings in the cortex, arrive at the medullary portion, where each pursues a straight course, and is lined with flat pavement-epithelium similar to the endothelium of vascular tissue.

Near the base of the pyramids these tubes curve upwards, forming the looped tubes of Henle. The recurrent tubes enlarge, and exhibit the ordinary cubical gland-cell. These tubes also become more tortuous, and empty into others of larger calibre, called collecting tubes. These are lined with low columnar epithelium, and uniting with similar tubes at acute angles, give exit to the urine at the apex of the papillæ in the pyramids (Plate XXIII, Fig. 170).

The bloodvessels of the kidney are as complex as the glandular tissue. Both vein and artery enter at the hilus of the kidney, and after giving twigs to the external tunic, proceed between the pyramids as far as their bases. Here they give off curving branches, forming imperfect arches among the arteries, and complete anastomosing rings on the veins. From the arterial arches spring the branches which bear the *glomeruli* of the cortical substance or Malpighian tufts (Plate XXIII, *a*, Fig. 171). The afferent vessel of the glomerulus subdivides, and after coiling and twisting within the capsule of Bowman, gives origin to the efferent vessel, by the union of the small branches thus formed. This efferent vessel breaks up into a network of fine capillaries, with elongated meshes surrounding the straight uriniferous canals. From the periphery of this network somewhat wider tubes are given off, which surround with rounded meshes the convoluted tubes of the cortex.

The long bundles of vessels between the uriniferous tubes of the medulla, communicating in loops or forming a delicate network round the mouths of the canals at the apex of the papillæ are called the *vasa recta*.

The *ureters*, like the pelvis of the kidney, consist of an external fibrous tunic, a middle layer of smooth muscular fibres, and an internal mucous membrane with a layer of epithelium. The *bladder* is covered externally with a serous membrane, the peritoneum. The female *urethra* is lined by mucous membrane, with vascular walls full of folds, and containing, near the bladder, a number of mucous glands.

3. *Respiratory Organs*.—The lungs receive air by the trachea and venous blood from the right side of the heart to transmit to the left side. They may be compared, as to form and development, to racemose glands. The excretory ducts are represented by the bronchial ramifications, and the acini by the air-vesicles.

The ciliated mucous membrane of the bronchial twigs gradually loses its laminated structure until only a single layer remains. Their muscular layer also ceases before arriving at the air-cells. At the end of the last bronchial tubules we find thin-walled canals called *alveolar passages*. These are again subdivided and end in peculiar dilatations called primary pulmonary lobules, or *infundibula* (Plate XXIII, Fig. 172). The *air-cells*, *vesicles*, or *alveoli*, are saccular dilatations in the walls of the primary lobules, opening directly into a common cavity. Their walls consist of delicate membrane of connective tissue, often containing black pigment, probably from inhalation of carbonaceous matter, or a deposit of melanin.

The pulmonary artery subdivides, and follows the ramifications of the bronchi to the pulmonary vesicles. Here a multitude of capillary tubes form a network over the alveoli, only separated from the air by the most delicate membrane (Plate XXII, Fig. 168). In the frog we find the whole respiratory portion lined with a continuous layer of flattened epithelia. A similar lining is found in the mammalian fœtus, but in the adult the number and character of the epithelial scales is greatly changed. Large non-nucleated plates are seen with occasional traces of the original bioplasm. In inflammatory affections, however, these may multiply, giving rise to catarrhal desquamation.

4. *Generative Organs*.—The histology of the organs of reproduction is quite elaborate, and the plan of this work

only permits us to glance at the essential structures, which are the seminiferous tubules for the secretion of spermatozoa, in the male, and the ovary for the production of the germ, or ovum, in the female.

The *tubuli seminiferi* are a multitude of fine and tortuous tubules contained in the testis, with its accessory epididymis. They lie in the interstices of sustentacular connective tissue, and consist of membranous tubes filled with cells, which are said to possess amœboid motion. During the virile period these glandular tubes generate the spermatozoa, or microscopic seminal filaments. The shape of these spermatozoa is filiform in all animals, but vary in different species. In man they consist of an anterior oval portion, or head, and a posterior flexible filament, or tail. Different observers have taken different views as to the origin of these structures. Some suppose them the product of special cells, others trace them to the nuclei of the glandular epithelium, while others regard them as ciliated elements formed by the metamorphosis of entire cells. Their motions baffle all attempts at explanation, although quite similar to those of ciliated epithelium. The spermatozoa penetrate by their movements into the interior of the ovum, in order to impregnate it, and in the mammalia in considerable numbers.

The *ovary* may be divided into two portions: a medullary substance, which is a non-glandular and very vascular connective tissue, and a glandular parenchyma enveloping the latter. The surface of the ovary uncovered by peritoneum is coated with a layer of low columnar cells, called the germinal epithelium. Immediately under this is a stratum called the zone of the primordial follicles, or cortical zone. Here the young ova lie crowded in layers. They consist of granular bioplasm, containing fatty molecules and a spherical nucleus. They are probably developed by a folding in of the germinal epithelium. Toward the internal portion of the ovary the follicles become more highly developed, and the ovum contained in them is also increased in size and enveloped in a distinct membrane. There are from twelve to twenty mature follicles in the ovarium, named, from their discoverer, Graafian follicles. Each has an epithelial lining, in which the ovum is imbedded. The capsule of the ovum is known as the *zona pellucida*, or *chorion*, and the albuminous cell-body is the *vitellus*. The nucleus is situated excentrically, and is called the *vesicula germinativa*, or germinal vesicle of Purkinje. Within it is a round and highly refractive nucleolus, the *macula germinativa*, or germinal spot of Wagner. A Graafian vesicle bursts and an ovum is liberated at every menstrual period. During the progress of the latter down the Fallopian tube to the uterus, impregnation may take place by the penetration of spermatozoa into its yelk. Then the inherent vital energies of the cell are aroused, and the process of segmentation begins. Unimpregnated ova are destroyed by solution. The ruptured and emptied Graafian vesicle becomes filled up with cicatricial connective tissue, which constitutes what is called the *corpus luteum*, after which it gradually disappears.

II. Organs of Animal Life.

1. *Locomotive.*—The microscopic structure of bone and muscle has been described in connection with elementary tissues. Tendons and fascias belong to the connective tissues.

2. *Sensory.*—The nervous apparatus of the body, whose histological elements were treated of on a previous page, has been classified physiologically into:

1. *The sympathetic system*, consisting of a chain of ganglia on each side of the vertebral column, with communicating cords or extensions of ganglia, visceral nerves, arterial nerves, and nerves of communication with the cerebral and spinal nerves. The chief structural differ-

ence between this and the cerebro-spinal system is that in the latter the nerve-cells form large masses, and the union of its parts is effected by means of central fibres, while in the sympathetic the cells are more widely separated, and union between them and with the cerebro-spinal axis is by means of peripheral fibres. The sympathetic is considered a motor and sensitive nerve to internal viscera, and to govern the actions of bloodvessels and glands.

2. *The cerebro-spinal system*, divided into:

(1.) A system of ganglia subservient to reflex actions, the most important of which is the spinal cord, where the gray or vesicular nervous matter forms a continuous tract internally.

(2.) A ganglionic centre for respiration, mastication, deglutition, etc., with a series of ganglia in connection with the organs of special sense: the medulla oblongata, with its neighboring structures; the mesocephalon, corpora striata, and optic thalami.

(3.) The cerebellum, a sort of offshoot from the upper extremity of the medulla, for adjusting and combining voluntary motions.

(4.) The cerebrum, cerebral hemispheres, or ganglia, which are regarded as the principal organs of voluntary movements. In the lower vertebrates the hemispheres are comparatively small, so as not to overlap the other divisions of the brain; but in the higher Mammalia they extend over the olfactory lobes and backward over the optic lobes and cerebellum, so as to cover these parts, while they also extend downward toward the base of the brain. In the lower vertebrates, also, the surface of the hemispheres is smooth, while in the higher it is complicated by ridges and furrows.

(5.) The cerebral and spinal nerves. The spinal nerves arise in pairs, generally corresponding with the vertebræ. Each has two roots, one from the dorsal, and one from the ventral region of its half of the cord. The former root has a ganglionic enlargement, and contains only sensory fibres; the latter has no ganglion, and contains only motor fibres.

The cerebral nerves are those given off from the base of the brain. Some of these minister to special sensation, as the olfactory, optic, auditory, part of the glosso-pharyngeal, and the lingual branch of the trifacial nerves. Some are nerves of motion, as the motor oculi, patheticus, part of the third branch of the fifth pair, the abducens, the facial and the hypoglossal nerves. Others are nerves of common sensation, as the fifth, and part of the glosso-pharyngeal nerves. Others, again, are mixed, as the pneumogastric and spinal accessory nerves.

The minute structure of the central organs of the nervous system is excessively complicate and full of details. Hardening with chromic acid and bichromate of potash is generally advisable before examination. This should be done with small pieces in a large quantity of the fluid. One-eighth to one-half grain of bichromate, or 0.033 to 0.1 grain of chromic acid, to the ounce of water should be used, the strength gradually increased from day to day. After such maceration for several days, a drop of a 28 per cent. solution of caustic potash may be added to one ounce of water, and the specimen soaked in it for an hour, to macerate the connective tissue. After again soaking in graduated solutions of the bichromate, up to two grains to the ounce, the tissue may be carefully picked apart under the dissecting microscope. In such manner Deiters discovered the two kinds of processes in the multipolar ganglion-cells. Gerlach placed thin sections for two or three days in 0.01 to 0.02 per cent. solutions of bichromate of ammonia, and picked them apart after staining with carmine.

Lockhart Clarke placed parts of the spinal cord in equal parts of alcohol and water for a day, then for several days in pure alcohol, till thin sections could be made. These were immersed for an hour or two in a mixture of one part acetic acid and three parts alcohol, to render the gray matter transparent and the fibrous elements prominent.

Sections may be stained with carmine and mounted in glycerin or balsam (see Chapter V).

(6.) Organs of special sense:

a. Organs of Touch.—The tactile papillæ of the skin and Pacinian corpuscles may be studied in thin sections of fresh or dried skin. Treatment with dilute acetic acid, or acetic acid and alcohol, and staining with carmine, or chloride of gold, is recommended. The papillæ are made up of connective tissue, into which nervous filaments enter, and end in peculiar tactile corpuscles (Plate XXIII, Fig. 173). The structure of the skin itself, with its various layers and sudoriparous glands, may be seen in such sections.

b. Organs of Taste.—The terminations of the gustatory nerves of the tongue are yet imperfectly known. In the circumvallate papillæ, on the side walls, certain structures are found, called *gustatory buds* or taste-cups (Plate XXIII, Fig. 174). They consist of flattened lanceolate-cells, arranged like the leaves of a flower-bud, and containing within them fusiform *gustatory cells*, which end in rods, and filaments projecting from the rods above the buds are seen in some animals. Underneath is a plexus of pale and medullated nerve-fibres. The mode of nervous termination in the fungiform papillæ is not known. For primary examination, sections of the dried tongue may be softened in dilute acetic acid and glycerin, or hardened in osmic acid. For the finer structure, maceration in iodine serum, and immersion in one-half per cent. chromic acid, with an equal quantity of glycerin, is recommended. Careful picking under the simple microscope is necessary. Sections may also be stained with chloride of gold.

c. Organs of Smell.—In the olfactory regions, which are patches of yellowish or brownish color on the upper and deeper part of the nasal cavity, we find nucleated cylindrical cells taking the place of ordinary ciliated epithelium, and sending processes downward, which communicate with each other, forming a delicate network (Plate XXIV, Fig. 175). Between these cells we find the olfactory cells, spindle-shaped nucleated bodies, extending upward into a fine rod and downward into a varicose filament. In birds and amphibia these rods are terminated by delicate hairs, some of which have ciliary motion. Beneath these structures are peculiar glands, consisting of pigmented gland-cells. They are called Bowman's glands. The branches of the olfactory nerve proceed between these glands and branch out into fine varicose filaments, which are supposed to communicate with the olfactory cells. Hardening in chromic acid, or Muller's fluid, or a concentrated solution of oxalic acid, or one-half to one per cent. solution of sulphuric acid, is necessary for the preservation of these delicate structures.

d. Organs of Sight.—As in the sense of touch certain tactile papillæ detect deviations from the general surface; and in that of taste special rod-like end organs and their covering bulbs distinguish the solutions of different sapid substances; and as in smelling, not the whole organ but olfactory regions, with peculiar cells and nervous rods, discriminate mechanical or chemical odors, so in vision a special apparatus is provided to perceive the wonderful variety of colors and forms. The minute structure of organs becomes more complex in proportion as they serve the higher functions of mind.

The various tunics and accessory structures of the eye are described in most text-books; we here limit ourselves to a brief reference to those refracting and receptive structures whose office it is to translate the phenomena of light into those of nervous conduction.

Externally, we have in front of the eye the transparent *cornea*. This is made of connective tissue with cells, bundles of fibres, and cavities containing cells. Its tissues are in layers, as follows: 1. External epithelium, flat and laminated. 2. Anterior basement-membrane or lamina. 3. True corneal tissue. 4. Membrane of Descemet or Demours. 5. Endothelium with flat cells (Plate XXIV, Fig. 176). The cells of corneal tissue are of two forms. The first are wandering or amœboid cells, and may be seen in a freshly extirpated frog's cornea placed underside up, with aqueous humor in a moist chamber, on the stage of the microscope. If a small incision be made at the margin of the cornea of a living frog a few hours before its extraction, and vermilion, carmine, or anilin blue is rubbed in, the cells which have absorbed the coloring matter will be found at some distance afterwards, having wandered like leucocytes or pus-cells elsewhere. Their origin may be from blood or true corneal corpuscles, or both. The second form, or corneal corpuscles, are immovable, flat, with branching or stellate processes. They may be demonstrated by staining with chloride of gold or nitrate of silver. The bundles of fibrillar substance in the cornea pass in various directions, and the natural cavities in it contain the corneal cells. As stated, the nerves of the cornea have been traced to the external epithelium, which sometimes contains serrated (riff or stachell) cells.

The *aqueous humor* is structureless, but the *vitreous humor* is supposed to have delicate membranous septa. The *crystalline lens* consists of a capsule inclosing a tissue of fine transparent fibres or tubules, which are of epithelial origin. These fibres are flat, and often have serrated borders, especially in fishes.

The *retina*, or nervous portion of the eye, is the most important, as its delicacy and liability to decomposition render it the most difficult object of microscopic examination.

We must dismiss the popular notion of minute images

produced on the retina by the lens to be viewed by the mind. The lens does, indeed, form an image on the membrane, so it would on glass or paper, but the real action of the vibrations of light upon the nervous conductors is not thus to be explained.

The complex structure of the retina is only recently known, and it may be that many laws of light yet unknown are to be exhibited by its means, as well as much that relates to the connection of the perceiving thinking mind and the external world.

Muller's fluid, concentrated solution of oxalic acid, 0.6 per cent. solution of sulphuric acid, and 0.1 to 2 per cent. solutions of osmic acid, may be used for hardening, but very delicate dissection is required for demonstration. Rutherford recommends chromic acid and spirit solution, 1 gramme of chromic acid in 20 c.c. of water, and 180 c.c. of methylated spirit added slowly.

The retina consists of the following layers: 1. The columnar layer, or layer of rods and cones. 2. Membrana limitans externa. 3. External granular layer. 4. Intergranular layer. 5. Internal granular layer. 6. Molecular layer. 7. Ganglionic cell layer. 8. Expansion of optic nerve. 9. Membrana limitans interna. To these may be added: 10. The pigment layer, often described as the pigmented epithelium of the choroid, into which the rods and cones project. These layers are composed of two different elements, mutually blended, a connective-tissue framework of varying structure in the different layers, and a complex nervous tissue of fibres, ganglia, rods, and cones. Plate XXIV, Fig. 177, is a diagram of these separate structures, after M. Shultze, in Stricker's *Manual of Histology*.

The structure of the rods and cones is complex, and varies in different animals. The rods readily decompose, becoming bent and separated into disks, but examination of well-preserved specimens shows them to have a fibrillated outer covering. In addition, certain globular or lenticular refractive bodies, of different shape and color in different animals, are found in these structures (Plate XXIV, Fig. 178), which doubtless are designed to give the rays of light such a direction for final elaboration in the outer segment as they could not receive from the coarser refractive apparatus in the front of the eye.

c. Organs of Hearing.—These are most intimately connected with mental functions, because of language, which is the highest sensual expression of mind. Hence the structure of these organs is most delicate and complex.

The labyrinth is the essential part of the organ, consisting in man of the vestibule, the semicircular canals, and the cochlea. Sonorous undulations are propagated to the fluid in the labyrinth through the tympanum and chain of otic bones.

The auditory nerves are distributed to the ampullæ and sacculi of the vestibule, and to the spiral plate of the cochlea. At the terminal filaments in the sac of the vestibule, crystals, called *otoliths*, of shapes differing in various animals, are inclosed in membrane. Hasse considers them to be vibrating organs, but Waldeyer regards their function to be that of dampening sound.

As we distinguish in sounds the various qualities of pitch, intensity, quality, and direction, it is probable that there is a special apparatus for each, but histology has not yet established this fully. Kölliker thinks the ganglionic termination of the cochlear nerve renders it probable that it only receives sonorous undulations. The experiments of Flourens seem to show that the semicircular canals influence the impression of direction of sound.

In the sacs of the vestibule and ampullæ, the nerve-fibres are confined to a projection of the walls called the septum nerveum. Here are found cylinder- and fibre-cells, with rods, basal-cells, and nerves. But it is in the lamina spiralis of the cochlea that the most elaborate organ,

PLATE XXIV.

Fig. 175.

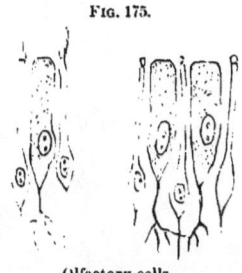

Olfactory cells.

Fig. 176.

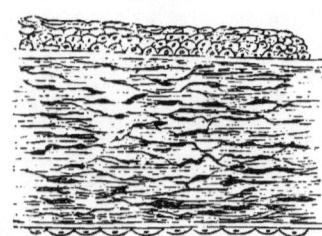

Section of Cornea.

Fig. 177.

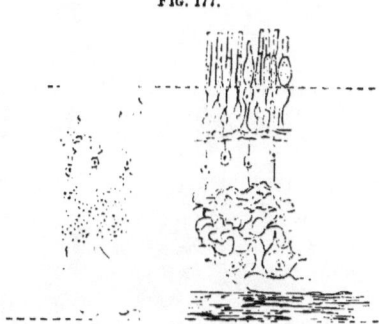

Connective-tissue and nerve-elements of Retina. Showing rods and cones.

Fig. 179.

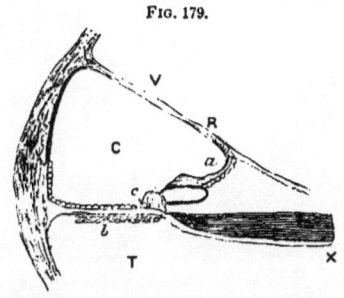

Section of Cochlea:—v, scala vestibuli; T, scala tympani; c, canal of Cochlea; R, Reissner's membrane, attached at a to the habenula sulcata; b, connective-tissue layer; c, organ of Corti.

Fig. 178.

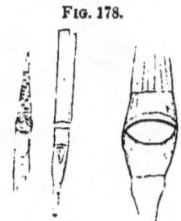

Refractive bodies in the rods and cones.

Fig. 180.

Corti's organ, from above.

Fig. 181.

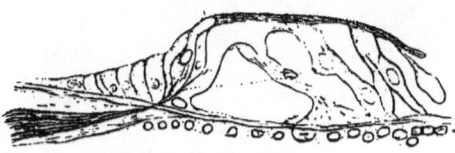

Section of Corti's organ.

called from its discoverer the *organ of Corti*, is found. Kölliker considers the free position of the expanded portion of the nerve, and the extent of surface over which its terminal fibres are spread, to constitute it an organ of great delicacy, enabling us to distinguish several sounds at once and to determine their pitch. There is a striking analogy between the visual and auditory apparatus in the ganglionic structure of the nerve-structure. Plate XXIV, Fig. 179, represents a vertical section through the tube of the cochlea; and Plate XXIV, Figs. 180 and 181, the vestibular aspect and a vertical section of Corti's organ.

Waldeyer recommends examination of the cochlea in a fresh state and in aqueous humor. Preparations in osmic acid and chloride of gold are also useful. For sections he removes much of the bony substance of large cochleæ with cutting pliers, opens the membrane in several places, and places the specimen in 0.001 per cent. of chloride of palladium, or 0.2 to 1 per cent. osmic acid solution for twenty-four hours, then for the same time in absolute alcohol. It is then treated with a fluid composed of 0.001 per cent. chloride of palladium with one-tenth part of $\frac{1}{4}$ to 1 per cent. muriatic or chromic acid, to deprive it of earthy salts. It is then washed in absolute alcohol, and inclosed in a piece of marrow or liver, and placed to harden in alcohol again. The hollows of the cochlea may be filled with equal parts of gelatin and glycerin before they are inclosed. Sections must be cut with a sharp knife.

Rutherford advises the softening of the bone and hardening of other tissues by maceration in chromic acid and spirit (1 gramme of chromic acid in 20 c.c. of water, and 180 c.c. of methylated spirit slowly added). For sections he commends Stricker's mode of imbedding in gum. Place the cochlea in a small cone of bibulous paper, containing a strong solution of gum arabic, for four or five hours; then immerse the cone in methylated spirit for forty-eight hours, or until the gum is hard enough. The sections may be stained with carmine, logwood, silver, or gold.

The following suggestions from Rutherford's *Outlines of Practical Histology*, will be of service to the student in this department:

Most of the tissues required may be obtained from the cat or guinea-pig. Feed the cat, and an hour or so after place it in a bag; drop chloroform over its nose until it is insensible. Open the chest by a linear incision through the sternum, and allow the animal to bleed to death from a cut in the right ventricle.

Divide the trachea below the cricoid cartilage and inject it with ¼ per cent. chromic acid fluid; tie it to prevent the escape of fluid, and place the distended lungs in the same fluid, and cover them with cotton-wool. Change the fluid at the end of eighteen hours. Allow them to remain in this fluid for a month, then transfer to methylated spirit till needed for mounting.

Open by a linear incision the œsophagus, stomach, large and small intestines, and wash them with salt solution (¾ per cent.). Place a portion of small intestine in chromic and bichromate fluid (1 gramme chromic acid and 2 grammes potassium bichromate in 1200 c.c. water) for two weeks (change the fluid at the end of eighteen hours), and then in methylated spirit till required. Act similarly with parts of œsophagus, stomach and large intestine, in ¼ per cent. chromic acid for three or four weeks. A portion of stomach may be placed in Muller's fluid till required for preparation of non-striped muscle, and of the gastric follicles.

The bladder may be treated as the small intestine.

Divide one kidney longitudinally, and the other transversely, and place in Muller's fluid. Change the fluid in eighteen hours, and after four weeks transfer to methylated spirits. They will be ready for use in two weeks after.

Cut one-half of the liver into small pieces and prepare as the kidneys. The tongue, divided transversely into five or six pieces, the spleen, uterus, and thin muscles from limbs or abdomen, in ¼ per cent. chromic acid. Change as before, and in a month to methylated spirit.

Testis of dog, freely incised, and ovaries of cat or dog, in Muller's fluid, and after three weeks to methylated spirits.

Divide the eyes transversely behind the lens. Remove the vitreous. Place posterior halves in chromic and spirit solution. Change in eighteen hours. Transfer to methylated spirit in ten days. Place the lens in Muller's fluid for five weeks, and then in methylated spirits. The cornea may remain in ¼ per cent. chromic acid for a month, and then in methylated spirit.

Cautiously open the cranial and spinal cavities. Remove brain and cord, and strip off arachnoid. Partially divide the cord into pieces a half inch long. Partially divide the brain transversely into a number of pieces. Place in a cool place in methylated spirits for eighteen hours. Transfer cord to ¼ per cent. chromic acid for six or seven weeks. Change in eighteen hours. Prepare the sciatic nerve in the same manner. Place the brain in chromic and bichromate fluid. Change in eighteen hours, and then once a week, until the brain is hard. If not leathery in six weeks place in ⅙ per cent. chromic acid for two weeks, and then in methylated spirits. Support the brain and cord on cotton-wool in the hardening fluid.

Remove muscles, but not periosteum from bones of limbs, and both from the lower jaw. Divide the bones transversely in two or three places, and put them in chromic and nitric fluid (chromic acid, 1 gramme; water, 200 c.c.; then add 2 c.c. nitric acid). Change the fluid often until the bone is soft enough, and transfer to methylated spirits. If not complete in a month, double the quantity of nitric acid in the fluid.

Place a piece of human scalp, skin from palmar surface of finger, and skin of dog (for muscles of hair-follicles) in chromic and spirit fluid. In a month transfer to methylated spirit.

Remove the petrous portion of temporal bone, open the tympanum, pull the stapes from the oval fenestra, and place the cochlea in chromic and spirit fluid. Change in eighteen hours, and at the end of seven days, if a brown precipitate falls, change fluid every third day. On the tenth or twelfth day transfer to chromic and nitric fluid. Change frequently till the bone is soft. Then place it in methylated spirit. The cochlea of the guinea pig projects into the tympanum, and is, therefore, convenient for enabling the student to see how the cone is to be sliced when sections are made.

Too long exposure to chromic acid renders tissues friable, and prevents staining with carmine.

Methylated spirit is ordinary alcohol containing 10 per cent. of wood-naphtha, and is used in England as a substitute for alcohol, since it is free of duty for manufacturing purposes.

CHAPTER XIII.

THE MICROSCOPE IN PATHOLOGY AND PRACTICAL MEDICINE.

PATHOLOGICAL HISTOLOGY, though yet imperfect, has attained an extended literature. Paget, Jones, Sieveking, Rokitansky, Virchow, Rindfleisch, and Billroth are names of investigators in this department, well known to medical students. As in our former chapters, we aim only to present the briefest and most elementary outline of the subject as introductory to more extended research.

I. Microscopic Appearances after Death of the Tissues, or Necrosis.

1. *Blood.*—This undergoes decomposition more rapidly than other tissues. The colorless corpuscle or bioplast, after slightly swelling, dissolves, and entirely disappears. The coloring matter leaves the red corpuscles shortly after death, and is diffused through the tissues, then the corpuscle disintegrates, and breaks up into granules.

2. *Nucleated Cells.*—In these the protoplasm coagulates, forming a solid albuminate, which becomes cloudy, and breaks up into granules.

3. *Cell-membrane* resists decomposition in proportion as it has become horny. Hence the outer layers of epithelium last longer than the inner ones.

4. *Smooth muscle fibres* are first filled with dusty particles, which unite into elongated masses, then they assume a striated appearance, and finally soften into a slimy matter.

5. *Striated Muscular Fibre.*—The muscle-juice coagulates to a solid albuminate, giving rise to rigor mortis in twelve or fourteen hours after death, except in death from charcoal or sulphuretted hydrogen vapor, lightning, or from putrid fevers or long debility. This stiffness of the muscle lasts about twenty-four hours. Under the microscope the transverse striæ and nuclei first disappear, then fat- and pigment-granules show themselves, the fibres melt away from the edges and became gelatinous. If gelatinous softening is marked, the fibres may disintegrate into Bowman's disks.

6. *Nerve-tissue.*—Little is known of its necrosis, beyond the fact that the white substance of Schwann first coagulates, then there is a collection of drops of myelin within the neurilemma, producing varicosity before complete dissolution.

7. *Adipose Tissue.*—The fluid fat leaves the cells and gives an appearance of emulsion to the mass.

8. *Fibres of loose connective tissue* swell, become stained with the coloring matter of blood, granulate, and liquefy, or they may desiccate by evaporation.

9. *Elastic fibres and networks* resist longer than the last. Hence elastic fibres may be found in expectorated matter from gangrene of the lungs, etc. Later, they break into granular striæ, then into molecules, and vanish.

10. *Cartilage* resists long, but melts away at the edge, first becoming transparent and reddish. The cells fill with fat-globules from fatty degeneration of bioplasm.

11. *Bone* retains its structure in necrosis, and hence is recognized by the surgeon in sequestræ, yet it decays in patches. The bioplasm becomes flat in the cells, acid fluids dissolve the lime salts, and the remaining structure disintegrates like cartilage.

The ichorous fluid into which all tissues are resolved, finally ends in carbonic acid, ammonia, and water, but the metamorphic substances resulting from the various preceding changes are not yet well known to histo-chemistry. Some of them are volatile, sometimes giving rise to emphysematous or crepitant gangrene, and are of bad odor; others are more solid, and produce interesting microscopic objects, as leucin, tyrosin, margarin, ammonio-magnesian phosphate, and pigment-granules. At page 135 we have referred to the presence of bacteria and vibriones, and their origin from fungi, in decaying animal matters.

II. Morbid Action in Tissues.

1. *Infiltration.*—This consists in the deposition or filtration of material from the blood, and is caused by adulteration of the blood or certain peculiarities of tissues. Thus for infiltration of fat, the liver and areolar connective tissue is most fitted; for superfluous salts of lime, the

lungs; amyloid matter seeks first the kidneys, next the spleen, liver, etc. In addition, there may be local causes, as pigment deposits from hyperæmia, hæmorrhage, etc.

(1.) Amyloid infiltration of tissues. This is a waxy, lardaceous, or vitreous albuminate, but may be distinguished from fibrin, albumen, etc., by becoming blue, violet, or red, with iodine. Sometimes it has concentric layers. Its likeness to starch led Virchow to call it amyloid.

Deposits of fibrin in blood extravasations in the lungs show a change into amyloid, round, small heaps of blood-globules, fragments of tissue, particles of charcoal, etc.

An amyloid infiltrated cell is larger than natural, and deformed, often coalescing with others.

The small arteries and capillaries, as the Malpighian tufts of the kidney, etc., are generally the first to suffer infiltration, which extends to the outer coat and surrounding tissue of the artery, even obliterating it. The degeneration of vessels leads to anæmia, as in lardaceous liver.

(2.) Calcification is the infiltration of tissue with solid phosphate and carbonate of lime. Free carbonic acid is the solvent of these salts, and by its capacity for diffusion it escapes, leaving the insoluble salts in the stagnating nutritive fluid. Thus cartilage becomes bone, and under peculiar circumstances other tissues calcify, as the pleuritic false membrane, etc.

(3.) Pigmentation. Under necrosis we referred to coloring matter of the blood impregnating tissues in soluble form; under this head we refer to it in solid form, as granules or particles, often without much depreciation of the functions of a part. The color of bile is derived from blood, and jaundice is an infiltration of fluid pigment from absorption of bile color. The black pigment of the lungs is from charcoal or inhaled carbon. Amœboid cells or leucocytes may imbibe solid particles and carry them in their wanderings.

(4.) Fatty infiltration. This is different from fatty metamorphosis, which is so common that infiltration may be presumed to be frequent. In fatty metamorphosis the globules are more numerous, but do not run together as in infiltration. The presence of soda compounds of bile in the blood, forming an emulsion with fat, may lead to infiltration, as chyme yields it in the intestine, or as liver-cells absorb it from the serum of the blood. Fat is sometimes removed from one place by metastasis to be deposited in another.

2. *Degeneration.*—(1.) Fatty degeneration is a metamorphosis of the bioplasm, marked by fat-globules in its interior. Thus in dropsy of the pericardium the epithelial cells first exhibit fat-globules, which by their aggregation in the albuminous matter enlarge the cell into a globular mass of granules. Gluge first called these "inflammatory corpuscles," but they are now known as fatty degenerated epithelium or granular corpuscles. These disintegrate to a fatty detritus. A large amount of granular corpuscles give the suspending fluid a yellowish color. The appearance of colostrum, on the first secretion of the mammæ, is due to this. By standing, it separates into a serous fluid and cream-like mass, the latter consisting of granular (colostrum) corpuscles.

The last act of fatty degeneration may be termed lactification. At the beginning of disintegration the Brunonian movement may be observed. Fats are finally partly saponified and partly separated in solid form, margarin and cholesterin.

In fatty degeneration of muscle we observe varicose fibrils or detritus, which render the striæ indistinct, or fill the sarcolemma with fluid.

Fatty metamorphosis is the regular mode of decomposition for tissues liable to rapid change, especially epithelium. Decreased nutrition may produce it, especially in non-vascular tissues, as in the cells of laryngeal cartilage, and in arcus senilis. Of vascular organs, the muscular tissue of the heart is most liable to it.

The cheesy degeneration of Virchow is a variety of fatty degeneration. It is a yellowish, compact, friable, or smeary mass, like cheese. It was formerly believed to be the product of tuberculosis, and regarded as the separation of morbid matter (crude tubercle) from diseased blood. It is now regarded as a fatty degeneration product, with less water present than usual. Sometimes salts of lime are infiltrated in such masses. Real tubercle is a gray, translucent, compact nodule, about the size of a millet-seed (miliary), found in great numbers together. Cheesy inflammation and miliary tuberculosis are often found side by side, and Cohnheim, etc., have shown that inoculation with cheesy detritus will produce tubercle. Tubercle is found in many organs, especially the lungs. Its structure consists of: 1. Large rounded cells of finely granular substance, and small strongly-shining nucleus or nuclei. 2. Small cells, with shining, darkly-contoured nuclei. 3. Mother-cells, with clear areas round the small ones. 4. A fine fibrous network (Plate XXV, Fig. 182).

(2.) Mucoid softening. Mucus is a colloid substance, capable of swelling by imbibition, but of little capacity for diffusion. It is a local production from epithelia of mucous membrane, yet is a structural element in many tumors. In the fœtus, the entire subcutaneous cellular tissue is mucous tissue. The fibrinous pseudo-membranes of the respiratory organs soften by mucoid metamorphosis, and in cartilage the intercellular substance dissolves in the same way, producing fibres near the surface.

(3.) Colloid degeneration. This is similar to the last, but differs in having peculiar cells, colloid globules, beginning in a normal cell by a change in its bioplasm, while mucoid softening occurs between the fibres of connective tissue. Like mucus, colloid enlarges by imbibition, and ends in soda albuminate, more soluble than common albumen, and identical with casein. Thus we have in the living and dead cell a circle of metamorphoses, from casein of milk to albumen of blood, thence to bioplasm, to formed material of cells and of intercellular substances, to mucus or colloid, and finally to casein.

III. New Formations.

To the surgeon the most important of these are tumors, excrescences, hypertrophies, or overgrowths. The nomenclature of such growths is, however, greatly redundant and often confusing. Some are named from the character of their contents as apparent to the eye, as hygroma (like water), melanoma (black pigment), chloroma (green ditto), hæmatoma (blood), colloma (glue), steatoma (lard), atheroma (gruel), meliceroma (honey), cholesteatoma (cholesterin), sarcoma (flesh), neuroma (nerve), encephaloma (brain), myeloma (marrow), schiroma (marble), etc.

Paget classified tumors as follows:
I. Innocent.
1. Cystic: Simple, compound, proliferous.
2. Solid: Fatty, fibro-cellular, fibrous, fibroid, cartilaginous, myeloid, osseous, glandular, and vascular.
II. Malignant: Infiltrating, ulcerating, multiplying.

Virchow's nomenclature is based on the divisions of hypertrophy, homeoplastic formations, and heteroplastic formations.

Histologically, the questions of origin and structure chiefly concern us. Such a study may yet lead to a true classification and rules of diagnosis. Virchow held that cells multiply by division at the place of the tumor, so that the newly-formed tissues substitute a certain amount of normal constituents. Cohnheim's wandering cells, however, show that formative elements may come from a distance, although local formation is also possible. Stricker shows in inflammation a division of both wandering and local cells. This favors Beale's view that multiplying bioplasm are the true disease germs.

The general plan on which new formations occur has been shown by Rindfleisch as: 1. The uniform enlargement of an organ by increase of structural elements (hypertrophy). 2. A node, or roundish tumefaction, by interstitial deposit, stretching the parenchyma. 3. Infiltration, which condenses tissue in small depots. 4. Desquamation, as in epithelial catarrh. 5. Flat tumefaction. 6. Tuberosity, which, when narrow or finger-like, is a papillæ or wart. 7. Fungus, spongy or ulcerated. 8. Polypus, a papilla with a narrow base. 9. Dendritic vegetation, or new papillæ from the sides of others. 10. Cysts of retention, from occlusion of ducts. 11. Exudation cysts, in closed cavities. 12. Etravasation cysts. 13. Softening cysts.

Histologically, we may divide tumors into:

I. Histoid, whose elements correspond with normal tissues.

II. Carcinomatous, dependent on abnormal growth of epithelial elements. These are generally in the skin, the mucous membranes, or glands.

The following list, after Billroth, may serve for brief reference:

1. Fibroma, composed of developed connective tissue. (1.) Soft fibrous tumors, almost exclusively in the skin. (2.) Firm fibroma. Most often in the uterus, where they may calcify or form fibrous polypi; sometimes on the periosteum and on nerves.

2. Lipoma, or fatty tumor.

3. Chondroma (cartilaginous). These occur on bones, are vascular, and often ossify. The softened and cystic forms have been called colloid tumors, gelatinous cancer, etc.

4. Osteoma (exostosis). These may be spongy or compact.

5. Myoma. Billroth doubts if true muscular tumors

exist. He objects to calling the spindle-celled fibroma of the uterus myoma.

6. Neuroma (nerve tumor). This also is misapplied to all tumors or nerves, many of which are fibroma.

7. Angioma (vascular tumor), as nævi and erectile tumors. They may be plexiform or cavernous.

8. Sarcoma, a large and uncertain group, containing (1) granulation sarcoma, or round-celled sarcoma of Virchow (Plate XXV, Fig. 183). (2.) Spindle-celled sarcoma (Plate XXV, Fig. 184). (3.) Giant-celled sarcoma (Plate XXV, Fig. 185). (4.) Net-celled or mucous sarcoma (Plate XXV, Fig. 186). (5.) Alveolar sarcoma, often resembling carcinoma (Plate XXV, Fig. 187). (6.) Pigmentary sarcoma, or melanoma (Plate XXV, Fig. 188).

9. Lymphoma, or enlarged lymph-glands.

10. Papilloma, or hypertrophied papillæ.

11. Adenoma, or glandular hypertrophy, as in goitre; sometimes forming mucous polypi.

12. Cystic tumors. If in connection with other tumors they are named accordingly, as cysto-fibroma, etc. They may be simple, compound, or proliferous.

13. Carcinoma, or cancers, so called from the distended veins sometimes appearing as crabs' feet.

The old division of cancers was into scirrhus or hard cancer, the outline well defined, the aspect of cut surface glistening, and yielding no juice; colloid, or soft cancer; encephaloid, or brain-like, yielding a milky juice full of cells and nuclei, generally soft, affecting neighboring glands, and associated with cancerous cachexia; fungus hæmatodes, when protruding and bleeding. Billroth treats of cancers (1) of skin, a glandular ingrowth of the rete Malpighi, and showing globular cells. (2.) Mammary cancers. (3.) Of mucous membranes with cylindric epithelium. (4.) Of lachrymal, salivary, and parotid glands. (5.) Of the thyroid gland and ovary.

It was thought by the early microscopists that a pecu-

PLATE XXV.

Fig. 182.

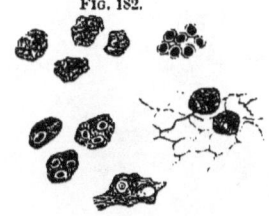

Tubercle.

Fig. 183.

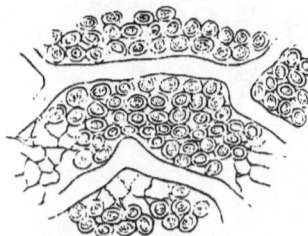

Granulation Sarcoma.

Fig. 184.

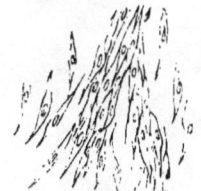

Spindle-celled Sarcoma.

Fig. 185.

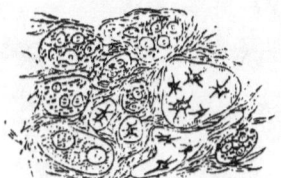

Giant-celled Sarcoma, with ossifying foci.

Fig. 186.

Net-celled Sarcoma.

Fig. 187.

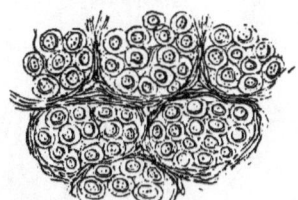

Alveolar Sarcoma.

Fig. 188.

Pigmentary Sarcoma.

Fig. 189.

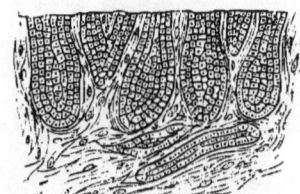

Epithelial Cancer of the Cheeks, Ingrowth of the rete Malpighi.

Fig. 190.

Epithelial Cancer of the Stomach.

Fig. 191.

Soft Glandular Cancer.

liar caudate cell was characteristic of cancer, but constant research has shown that there is no special form of cell in this disease. The greater number of cancers may be characterized as a proliferation and ingrowing of epithelial elements, either of the surface or of the glands. Figs. 189 to 191, Plate XXV, illustrate several varieties of carcinoma.

The anomalies of the various tissues and organs are well delineated by Rindfleisch in his *Pathological Histology*.

IV. MICROSCOPIC EXAMINATION OF URINARY DEPOSITS.

Healthy urine holds in solution a variety of organic and inorganic substances, as urea, uric acid, alkaline and earthy salts, animal extractive, vesical mucus, and epithelial débris. A few drops allowed to evaporate on a glass slide will exhibit the crystalline matters, consisting of urea, urate of soda, chloride of sodium, phosphates and sulphates.

The amount passed each twenty-four hours varies from 20 to 50 ounces, holding in solution from 600 to 700 grains of solid matter. Of this, about three-fourths consists of organic, and one-fourth of saline substances, the largest amount being urea, comprising nearly two-thirds of the whole. The amount, both of solids and fluids, is subject to great variation, according to the amount of fluids imbibed, the action of the skin, etc.

The average specific gravity of healthy urine is 1.020. It may be measured with the urinometer, a loaded glass bulb, with a graduated stem. According to a table calculated by Dr. G. Bird, after Dr. Christison's formula, each degree of the urinometer represents 2.33 grains of solids in 1000. Thus the specific gravity 1.020 represents 46.60 grains of solid matter in 1000 of urine. By weighing all the urine passed in twenty-four hours, it is easy, therefore, to calculate the amount of solids secreted.

Dr. Bird has also given another table, from which it appears that the figures of specific gravity will indicate nearly the amount of solids in each fluid ounce. Thus, specific gravity 1010 shows a little more than 10 grains of solids to the ounce; 1020 equals a little over 20 grains; above 1030 a grain or two more must be added, as 1030 equals $31\frac{1}{2}$ grains; 1035 gives about 37 grains.

The proportion of urinary excretion to the weight of the body is often an important consideration. It may be stated, as an average, to consist of about 149 grains of water and $6\frac{1}{2}$ grains of solids to each pound weight in twenty-four hours.

The tables given by different observers vary, but the above may serve as an average approximation.

Urea is the vehicle by which nearly all the nitrogen of the exhausted tissues is removed from the system, and its retention is often attended with fatal uræmic poisoning of the blood. In health, 400 to 500 grains are excreted in twenty-four hours, but in some cases of kidney diseases not more than 100 grains are eliminated, while in some fevers over 1000 grains are removed in the same period. If urea be suspected in excess, a drop of urine added to a drop of nitric acid may be placed under the microscope, when the characteristic crystals of nitrate of urea will appear (Plate XXVI, Fig. 192).

Volumetric analysis is the best means of ascertaining the quantity of urea, as of other chemical ingredients, but the practitioner may approximately estimate by weighing the crystals of nitrate of urea formed by adding nitric acid to double the quantity of urine, which has been concentrated to half its bulk by boiling.

The proportion of *uric acid* varies from 0.3 to 1 part in 1000 of healthy urine. It may be obtained by adding a few drops of hydrochloric acid to urine concentrated to half its bulk, and allowing it to stand in a cool place.

The estimation of the *chlorides* in urine is sometimes

necessary in disease. It may be done by acidulating the urine with a few drops of nitric acid, and then adding nitrate of silver. The precipitate should then be dried and fused in a porcelain capsule before weighing.

Albumen in suspected urine may be tested by boiling in a test-tube, when it will be coagulated. As a white precipitate sometimes occurs from an excess of earthy phosphates, a few drops of nitric acid should be added, which dissolves phosphates, but coagulates albumen.

Diabetic sugar is recognized by several tests. Moore's test is made by mixing the urine with half its bulk of liquor potassæ, and boiling gently for five minutes. Sugar gives the liquid a brown or bistre tint. Trommer's test consists in boiling the urine with a mixture of caustic potash and sulphate of copper, when if sugar be present the suboxide of copper will be reduced to a reddish-brown or ochre-colored powder. Fehling's test solution is a modification of the last, and is made by dissolving 69 grains of sulphate of copper in 345 grains of distilled water; to this is added a concentrated solution of 268 grains of tartrate of potash, and then a solution of 80 grains of carbonate of soda to 1 ounce of water, and the whole diluted to 1000 grains. The fermentation test consists in filling a test-tube with urine, to which a little yeast is added. The tube is then inverted over a saucer containing a little urine, and placed in a warm place for twenty-four hours. If sugar is present, it undergoes vinous fermentation, yielding alcohol and carbonic acid. The latter rises in the tube and displaces the liquid.

The *coloring matter of bile* in urine may be detected by the nitric-acid test. A few drops of biliary urine are poured on a white plate, and a drop of nitric acid allowed to fall upon it. As the acid mixes with the fluid, a play of colors, commencing in green, passing through various shades, and terminating in red, will be observed.

We necessarily omit many chemical details respecting

urinary examinations, but the foregoing will be found of practical use to the student.

Microscopic deposits in urine are either organic mixtures or precipitates from solution. They should be collected by standing several hours in a conical vessel.

1. *Organic Mixtures.*—(1.) Epithelium. The character of the scales will serve often to show the locality of disease in the urinary organs. The appearance of cells from various parts is shown in Fig. 142.

(2.) Mucus and pus-cells. Mucus is deposited as a flocculent cloud, entangling a few round or oval delicately granular cells, a little larger than a red blood-globule. In disease this increases and contains numerous ill-defined cells. A very thick glairy deposit in disease of the bladder may be mistaken for mucus, although it is pus altered by the action of carbonate of ammonia.

As pus is often formed from the germinal matter of epithelial cells, a small quantity in the urine is not necessarily a sign of serious disease. In large quantities, pus forms an opaque cream-colored deposit, which becomes glairy and tenacious on the addition of liquor potassæ. Pus-globules under the microscope, if long removed from the body, are granular, and show from one to four nuclei when treated with acetic acid. In fresh pus-corpuscles, especially in warm weather, amœboid motion is often seen. In a late period of catarrh of the bladder, but little epithelium will accompany the discharge, but crystals of triple phosphates generally occur in pus derived from the bladder.

(3.) Blood-disks usually form a reddish-brown deposit. A smoky appearance of the urine is produced in blood derived from the kidney. A brown deposit resembling blood, but showing granules and no disks, is supposed to be derived from blood. The epithelium associated with blood-disks will often point out the source of hæmorrhage.

(4.) Spermatozoa in urine are not uncommon in perfect

PLATE XXVI.

Fig. 192.

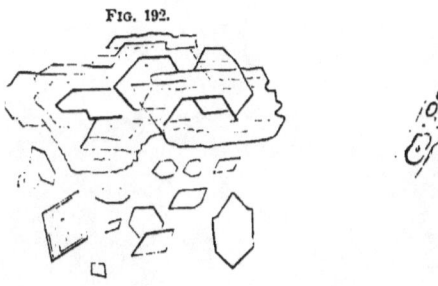

Nitrate of Urea.

Fig. 193.

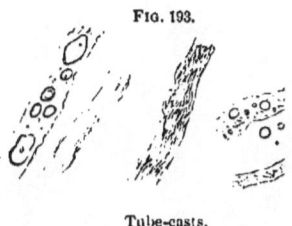

Tube-casts.

Fig. 194.

Urate of Ammonia.

Fig. 195.

Fig. 196.

Uric Acid.

Uric Acid.

Fig. 197.

health, but nervous patients are often deluded by quacks on account of them.

(5.) Accidental products. A great variety of things may accidentally get into urine, and the observer must guard against them by studying the appearance of various objects in the microscope. Pieces of feathers, fibres of wood, vegetable-cells, wool, cotton, silk, dust, etc., will almost always attract the attention first of one unused to them.

(6.) Sarcinæ are minute vegetable organisms, in the form of cubes, often subdividing into groups of four or their multiples. They were detected by Goodsir in the stomach in a case of obstinate vomiting, and have been occasionally found in urine. They seem associated with some dyspeptic cases.

(7.) *Torulæ*. This fungus is developed in urine which contains even minute traces of sugar. It is identical with the yeast plant (see pages 135, 137).

(8.) The *penicillium glaucum*, a fungus allied to the last, commonly makes its appearance in acid urine when exposed to the air.

(9.) Vibriones. What was said at page 135 will readily account for the presence of vibriones in decaying urine. In perfectly fresh urine it may be regarded as a sign of debility.

(10.) Tube-casts. In many cases of congestion and inflammation a coagulable material is effused into the tubes of the kidney, forming a cast or mould of the tube. This may be ejected, bringing with it pus, blood, epithelium, or other material with which it is associated. In Bright's disease these casts, in addition to albuminous urine, assume considerable clinical importance. In the acute form of the disease the cylinders or casts are fibrinous, with blood, mucus or pus-cells, and epithelium. Towards the close the casts become homogeneous or hyaline. In chronic desquamative nephritis the cylinders are without blood, and towards the close waxy or fatty, often containing many oil-globules (Plate XXVI, Fig. 193).

2. *Precipitates from Solution.*—(1.) Urate of ammonia. This is generally an amorphous deposit, in irregular groups of molecules, but with an alkaline fermentation sometimes crystallizes (Plate XXVI, Fig. 194). Some regard this as urate of soda, or a mixture of urates of potash, soda, and ammonia.

Its color varies from light pink to brickdust color. It is deposited in all concentrated urine, and is often a "critical discharge" in fevers, etc. It is found in gouty concretions, and dissolves with heat and acids.

(2.) Uric acid may occur from an acid fermentation dissolving urates of soda or ammonia. It is a yellow, reddish, or brown sediment of crystals, which assume different forms, as rhomboid tablets with obtuse angles, or of the shape of a whetstone (Plate XXVI, Fig. 195).

When slowly precipitated, it may form druses of four-sided prisms (Plate XXVI, Fig. 196). When precipitated from fresh urine by the addition of muriatic acid, the crystals are large, and often of varying shapes.

They may be tested by dissolving in potassa, and re-precipitating by muriatic acid, when they assume the shape of Fig. 197, Plate XXVI.

They originate from waste, excess of nitrogenous food, defective assimilation, congestions of the kidney, or chronic disease of the respiratory organs.

(3.) Ammonio-phosphate of magnesia, triple phosphate, may be precipitated from fresh urine in stellate crystals (Plate XXVII, Fig. 198) by adding ammonia. When more slowly deposited from alkaline urine, or in diseased states of the system, the crystals are prismatic, generally triangular, with obliquely truncated ends. Sometimes the terminal edges are bevelled, and the varying lengths of the prisms give rise to a variety of forms (Plate XXVII, Fig. 199). They are generally thought to proceed from

PLATE XXVII.

Fig. 198.

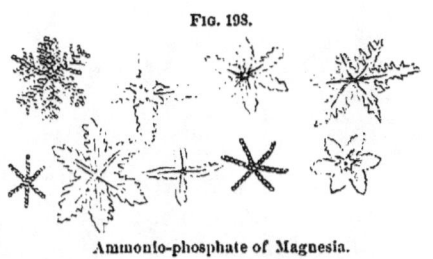

Ammonio-phosphate of Magnesia.

Fig. 199.

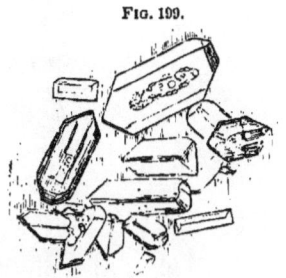

Ammonio-phosphate of Magnesia.

Fig. 200.

Phosphate of Lime.

Fig. 201.

Fig. 202.

Chloride of Sodium.

Oxalate of Lime.

Fig. 203.

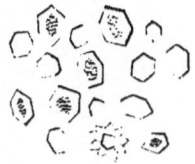

Cystine.

Fig. 204.

Salivary corpuscles, epithelial scales and granules.

disintegrated albuminous, and chiefly nervous, matter, but their clinical importance is not fully settled. They are found in cases of nervous depression, various forms of dyspepsia, shock of the spinal cord, irritation of the bladder, etc.

In highly alkaline urine, the triple phosphates are usually accompanied with pus and phosphate of lime. The latter occurs as minute granules or dumb-bells, or in groups of crystals (Plate XXVII, Fig. 200).

(4.) Oxalate of lime is deposited in small octahedra, generally appearing under the microscope as minute squares, with crossed lines proceeding from the angles, the upper angle being next the eye (Plate XXVII, Fig. 201). Dumb-bell forms, and circular or oval crystalline masses, are often seen.

Oxalate of lime is found as a urinary deposit in various conditions, as pulmonary and dyspeptic affections. It is usually associated with hypochondriasis, and in cases of overfatigue, particularly from mental work, it is very common. Its association with calculous affections renders it interesting to the surgeon.

(5.) Chloride of sodium never crystallizes from fluid urine. On evaporation it crystallizes in stellar form or in cubes (Plate XXVII, Fig. 202). The presence of urea sometimes disposes it to assume the form of a regular octahedron.

It is useful to investigate this excretion in typhoid fevers and inflammations of the respiratory organs, etc. In commencing hepatization of the lung it is absent, but returns on resolution of the inflammation. The method of testing has been given before.

(6.) Cystin (Plate XXVII, Fig. 203) crystallizes in characteristic six-sided plates. It contains a large proportion of sulphur, 26 per cent., and is considered a product of decomposition. It is often associated with calculus. Some regard it as indicating a strumous and ill-nourished system.

V. List of Parasites Infecting the Human Body.

I. *Epiphytes*, or vegetable parasites. Parasitic lesions of the surface are denoted by the infiltration or destruction of hairs and epithelial textures by the sporules of a fungus, which by union or growth form elongated branches or mycelia. Reference has been made, page 136, to polymorphism, or the varieties of form produced by the same fungus germ, so that the names ascribed to these parasites must be regarded as only provisional. The diagnosis of fungi on the skin, hair, or epithelium requires care and skill in microscopic manipulation, and the use of liquor potassæ long enough to render the specimen transparent.

1. The *Trichophyton tonsurans*, present in ring-worm of the body, scalp, or beard. Its anatomical seat is the interior of the roots of the hairs, but it also covers the epidermis between the hairs, and invests them in a white sheath, producing inflammation of the follicles and surrounding tissues, and subsequent baldness.

2. The *Trichophyton sporuloides*, present in the disease called plica polonica.

3. *Achorion Schönleinii* and the *Puccinia favi*, present in the honeycomb ring-worm.

4. *Microsporon mentagraphyta*, present in the Mentagra.

5. *Microsporon furfur*, the cause of liver-colored spots, or Pityriasis versicolor.

6. *Microsporon Audouini*, occurring in Porrigo decalvans, or bald patches.

7. *Mycetoma Carteri*, the cause of the "fungous foot of India."

8. *Oidium albicans*, in diphtheria and aphtha.

9. *Cryptococcus* (or Torula) *cerevisiæ*, yeast plant in bladder or stomach.

10. *Sarcina ventriculi*, in the stomach.

II. *Epizoa*, or animals living upon the skin and hair.

1. *Pediculus*, or louse, three forms: *P. corporis*, *P. capitis*, and *P. pubis*.
2. *Acarus scabiei*, or itch insect.
3. *Demodex folliculorum*, inhabiting sebaceous and hair-follicles.

III. *Entozoa*, or internal parasites. On page 171 we have given a general account of the Entozoa. At least thirty different forms have been described as infesting the human body; eight species of *Tænia* and two of *Bothriocephali*, genera of the family Cestoidea or tape-worms, of which the Cysticerci and Echinococci (vesicular cysts containing an embryo head provided with a circle of hooklets, and giving rise to the appearance of measly flesh in animals) are larval forms; nine species of Trematoda, or fluke-like parasites, existing in an encysted and a free state; and eleven species of Nematoid, or round-worms, including the common round-worms or *Ascaris*, the *Trichina spiralis*, and the *Filaria oculi*, etc.

Students who have not access to Dr. Cobbold's great work on parasites, may find an excellent résumé in Dr. Aitken's *Science and Practice of Medicine*.

VI. Examination of Sputa.

The microscopic examination of sputa is important in practical medicine, but requires familiarity with the appearance of different structures under different magnifying powers, as fragments of food, muscular fibre, starch, etc.

We may usually expect to find mucus, entangling air-bubbles, and pavement-epithelium from the mouth (Plate XXVII, Fig. 204). In catarrhal affections, ciliated epithelium from the nasal or respiratory passages may also be seen, and perhaps molecules of fat, pus-globules, blood, or transformed epithelial cells (formerly called granule-cells, or inflammatory corpuscles). In phthisis, the softened tubercle or gangrene may be early detected by the fibres of elastic tissue from the walls of the pulmonary vesicles. The sputa should be first liquefied by boiling with an equal bulk of caustic soda, and then allowed to settle in a conical glass, when a small quantity may be removed by a pipette to a glass slide, covered by thin glass, and placed under the microscope.

The occurrence of fungi in sputa is to be expected whenever there is decay. The *Leptothrix buccalis*, one form of *Penicillium*, is common on old epithelial scales of the mouth, and in the later stages of phthisis the sputa will often show fungi in different stages of development. Bacteria and vibriones are also frequent in pus.

In catarrhal pneumonia we may find fibrinous casts of the alveoli of the lungs and epithelial elements.

VII. Hints on the Application of the Microscope to Materia Medica and Pharmacy.

The observations of Dr. Hassall on the detection of adulterations in food,* have prompted similar investigations respecting the purity of medicinal substances. Such examinations cover a wide field of research, chiefly related to micro-chemistry and botany.

The student in this department will do well to provide himself with undoubted specimens of various articles for comparison, although much may be learned from a general examination of any particular drug, etc.

In addition to the recognition of genuine forms of leaves, seeds, roots, etc., and their adulterations, the microscope will often be serviceable in exhibiting the deteriorations to which such articles are subject if kept too long.

Dr. Hale, in the *American Journal of Microscopy*, shows

---

* Food and its Adulterations. By A. H. Hassall, M.D.

that various causes combine to effect the deterioration of drugs. They may become infested with animalculæ. Leaves and roots may be eaten by insects until all vestige of medicinal power is destroyed. Fungi of various kinds may destroy the tissue. The uncertain action of some pharmaceutical preparations may be thus accounted for.

For the method of examination in this department, as well as in medical jurisprudence, we refer to the foregoing chapters, contenting ourselves here with the remark, that investigations involving life or reputation should never be undertaken without a thorough practical acquaintance with microscopic manipulation and the microscopic sciences. He to whom such work is intrusted, should be able to exhibit and to explain to an intelligent jury what he has seen and what he ought to see.

GLOSSARY.

Aberration (*ab*, from, and *erro*, to wander).—The errors resulting from the imperfection of lenses. They are of two kinds, chromatic and spherical aberration.

Abiogenesis.—The dogma of spontaneous generation, or the alleged production of living beings without pre-existing germs.

Absorption Bands.—Transparent substances are usually opaque to certain colored rays of light, that is, absorb them; and when they are submitted to prismatic analysis, this opacity causes gaps in the spectrum. Some substances produce absorption lines of great sharpness, while others have an indistinct outline.

Achromatic.—Destitute of chromatic aberration.

Alternation of Generations, or *Metagenesis*, is a term employed to designate a cycle of phenomena in which one generation does not produce a form like itself, but one whose progeny are similar to the generation preceding.

Amœboid Movements.—Motion in minute masses of bioplasm, resembling that of *Amœba*. It has been recognized in Volvox, in Chara, in the roots of mosses, in fungi, as well as in colorless blood-cells.

Amyloid Infiltration—A filtration of waxy or lardaceous albuminate from the blood among certain tissues of the body.

Angular Aperture.—The angle made by the diameter of the actual aperture of an objective and the distance from its focal point. A very wide angle of aperture only allows distinct vision of what is exactly in focus, so that when penetration is needed most, as in physiological work, a smaller angle is better than one used for resolution of diatoms and other minutiæ.

Aplanatic.—Destitute of spherical aberration.

Archeus.—The term applied by Van Helmont to the specific agent which, according to his theory, presided over vital functions.

Bacteria.—Minute, transparent, rod-like bodies, sometimes jointed, and often exhibiting a vacillating motion. It is probable that they are produced by the germs of fungi in a solution of animal matter.

Bathybius.—A term given by Professor Huxley to the slimy matter from the ocean bottom. Doubts have been expressed as to its animal nature.

Binocular (binus, two, and *oculus,* an eye).—An arrangement of a prism (Wenham's) and eye-pieces for two eyes, so as to produce a stereoscopic effect with the microscope.

Biology (Gr. *Bios,* life, and *logos,* a discourse).—The study of living beings, including zoology and botany.

Bioplasm.—A term proposed for the elementary substance of organic bodies when actually alive, or living protoplasm.

Blastema.—The fluid matter, or plasma, in which, according to the theory of Schleiden and Schwann, nuclei first make their appearance, and then organic cells.

Brunonian Motion.—The molecular movement of fine particles suspended in fluid, first observed by Dr. R. Brown in 1827.

Calcification.—The infiltration of animal tissues with salts of lime.

Cell.—The elementary unit of organic structure. From the time of Schleiden and Schwann (1838) the researches of biologists have been greatly aided by the demonstration of the development of all living things from cells. A cell-wall, cell-contents, and a nucleus, were formerly regarded as essential, but further investigation has shown that a cell is essentially a semisolid mass of living matter.

Chlorophyll.—The green coloring matter of vegetables.

Chromatic Aberration.—The errors depending on the unequal refrangibility of the colored rays which make up white light.

Ciliary Motion.—The movement of cilia, or minute hair-like bodies, on animal and vegetable cells. The cause of it is unknown.

Colloid.—Substance devoid of crystalline power, as gum, albumen, gelatin, etc. Such substances pass slowly through a membrane, while crystalloid bodies pass readily, thus enabling us to separate them by dialysis.

Correlation of Forces.—The doctrine that any one of the forms of physical force may be converted into one or more of the other forms.

Cryptogamia.—Lower orders of plants, whose fructification does not depend on the presence of stamens and pistils in the flower.

Crystallography.—The science which treats of the laws by which the surfaces of crystals are disposed to one another.

Crystalloid.—Substances capable of crystallization.

Cyclosis.—A circulation of fluid in the cells of plants.

Definition.—The power of an object-glass to give a distinct image of an object.

Diaphragm.—A stop for intercepting some of the luminous rays, generally placed just beneath the microscope stage.

Diffraction of Light.—A disturbance of the straight path of a ray of light from its passage close to the edge of an opaque body.

Double Refraction.—The power possessed by some crystals, as Iceland spar, of exhibiting two images. The polariscope tests this property and exhibits it in many substances. A thin film of selenite intensifies it if it is not strong enough to show color otherwise.

Epiphytes.—Parasitic plants.

Epithelium.—The layer or layers of cells covering external or internal surfaces of animal bodies.

Epizoa.—Parasitic animals.

Focal Distance.—The distance from the centre of a lens to the focus, or point of distinct vision.

Foraminifera.—Shells of minute animals, chiefly calcareous, which are perforated with minute pores for the protrusion of threads of sarcode or bioplasm.

Formed Matter.—The structure produced by the action of bioplasm, or by the influence of external agents upon it.

Fraunhofer's Lines.—The dark lines which cross the solar spectrum, corresponding to the chemical nature of the burning substance. The blackness of the lines depends on the incandescent vapor of the substance.

Gemmation, or *Budding.*—A term given to the reproduction of cells by the protrusion of a part of their substance, which becoming constricted, falls off and lives an independent life.

Germinal Matter.—Another name for bioplasm, or "cell-stuff."

Herapathite.—The iodo-disulphate of quinia.

Histo-Chemistry.—The science which investigates the chemistry of the tissues.

Histology.—The science of tissues.

Immersion Lens.—An objective arranged so as to require a drop of fluid interposed between its front lens and the covering glass of the object.

Indifferent Fluids.—Fluids which produce little or no change in animal or other tissues.

Infiltration.—The deposition of material from the blood into various tissues.

Lens.—A piece of glass ground and polished so as to

refract the light to a focus, or causing the rays to diverge.

Leucocytes.—White cells, whether in blood or elsewhere. They are simply masses of bioplasm.

Luteine Spectra.—The spectra produced by light passing through juice from the corpora lutea in the ovary.

Metamorphosis.—Change of form, as from the caterpillar to the butterfly.

Metric Measure.—The system first adopted in France, based on the metre, which is the ten millionth of the quadrant of the meridian of Paris. The unit of surface is the *are* of one hundred square metres. The unit of weight is the gramme, weighing $\frac{1}{10000}$th of a cubic metre of water. The multiples are indicated by Greek prefixes, *deca* (10), *hecto* (100), *kilo* (1000), *myrio* (10,000). The subdivisions are named by Latin prefixes, *deci*, *centi*, and *milli*.

Micro-gonidia.—Bodies resulting from the segmentation of motile cells in the lower order of vegetables. When possessing active movement they rank as zoospores.

Micrometer.—An instrument for measuring minute spaces.

Microscopy.—The use of the microscope, and the knowledge attained by it.

Microzymes.—Minute molecules found in the vaccine vesicles, glanders, and other disease products.

Molecular Coalescence.—A name given to the action of various chemical substances, in a nascent state, upon an organic colloid.

Monera.—The name given by Professor Hackel to the simplest forms of animal life.

Morphology.—The science of form. Applied to the structures of organized beings.

Motile Cells.—Minute vegetable cells, moving by means of vibratile cilia. After a time they lose their cilia and become still cells, which multiply by self-division.

Necrosis.—The death of tissue.

Nucleus.—A concentration of vital power in a mass of bioplasm.

Objective.—The object-glass of a microscope.

Oblique Illumination.—The illumination of microscopic objects by light thrown from the side, either by the mirror or some special contrivance.

Oolites.—Certain rocks which present a granular structure resembling the roe of a fish.

Orbitolites.—Foraminiferous shells of considerable size occurring in tertiary limestones.

Otoliths.—Small crystalline bodies from the inner ear.

Pabulum.—The nutritive material supplied to animal cells.

Parthenogenesis.—Reproduction without sexual union.

Pathology.—The science of diseased structure and function.

Penetration.—The property of an objective which exhibits layers of structure below the focus. It seems to depend on moderate angular aperture.

Physical Movements.—A peculiar vibratile motion in minute particles suspended in fluid. See *Brunonian Motion.*

Polymorphism.—The development of similar germs into different forms by various agencies.

Protophytes.—Plants of simplest forms.

Protoplasm.—The elementary cell-material, or "physical basis of life."

Protozoa.—Simplest forms of animals.

Raphides.—Crystals occurring in vegetable tissues.

Resolution.—The property in a microscope of exhibiting minute details, as lines, etc.

Sarcode.—A synonym of protoplasm, or cell-material.

Sclerogen.—Woody tissue.

Selenite.—Crystallized sulphate of lime.

Spectrum Analysis.—The analysis of incandescent substances by means of the spectroscope.

Spherical Aberration.—The errors of lenses arising from the spherical surface, so that the rays from centre and edge do not accurately combine in focus.

Spontaneous Generation.—The theory of the spontaneous or independent origin of minute organisms. Bastian's views in favor of this theory seem to have been overthrown by the experiments of Pasteur and Tyndall, and biologists now generally agree to the doctrine that all living bodies are derived from pre-existent life.

Torula.—The "yeast-plant" fungus.

Vernier.—A short graduated scale, made to slide along a large scale so as to read to fractions of divisions.

Vibriones.—Vibratile filaments, or bacteria, sometimes moniliform, or bead-like, frequently found in decaying animal infusions.

Wandering Cells.—Leucocytes, or white blood-cells, which pass through the coats of vessels and thence into neighboring tissues.

INDEX.

Aberration of lenses, 22
Abiogenesis, 125
Absorption bands, 18, 101
Acalephs, 166
Acari, 177
Achromatic condenser, 33
Acinetœ, 162
Accessories, microscopic, 32
Adjustment, 51
Agriculture, microscope in, 18
Air-pump, 79
Alternation of generations, 126
Algæ, 139, 153
Albumen in urine, 237
Alimentary canal in insects, 177
Amici's prism, 34
Amœba, 121
Amœboid motion, 121
Amplifier, 26
Amyloid infiltration, 229
Amphlipleura pellucida, 56
Analysis of urine, 235
    of earths, etc., 99, 108
Anatomy of insects, 177
Animal histology, 182
Animalcule cage, 41
Angular aperture, 25, 55
    measurement of crystals, 89
Annular vessels, 131
Annulata, 172
Antennæ of insects, 175
Antiquity of microscope, 17
Aquaria, 82
Arachnida, 179
Aristotle on life, 116
Archeus, 116

Bacteria, 135
Bathybius, 96, 158
Beale's generalization, 118
    tint-glass camera, 40
Beck's microscope, 32
    iris diaphragm, 33
    illuminator, 38
Biology, microscope in, 116

Binocular microscope, 30
Bioplasm, 118
    varieties of, 123
Blood, 186
Blood-tests, 102
Blastema, 124
Bone, 195
Brunonian movement, 53, 120
Bryozoa, 168
Bull's-eye condenser, 36

Cabinet, 81
Carmine staining, 69
Care of microscope, 48
    of the eyes, 49
Calcification, 229
Camera lucida, 39
Camphor, 133
Cancer, 234
Capillary structure, 201
Cavities in crystals, 89
Cell, 117
    formation of, 119
    phenomena of, 120
    movements of, 120
    chemistry of, 122
    forms of, 123
    genesis, 124
    multiplication, 125
    wall, vegetable, 129
Cellulose, 129
Cements, 75
Cephalopoda, 171
Cerebro-spinal nerves, 216
Characeæ, 154
Chalk strata, 95
Chemical isolation, 59
    reagents, 67
    products, 183
Chlorophyll, 133
Chromatic aberration, 22
Chyle, 190
Cilia, 124
Cirrhipeds, 174
Classes of microscopes, 28

INDEX.

Ciliary motion, 161, 170
Chloride of sodium, 241
Coal, to prepare, 92
Colors of flowers, 133
Collecting objects, 81
Coddington lens, 23
Colloid, 66
    degeneration, 231
Compound microscope, 23
    crystals, 89
    tissues, 197
Collins's Harley microscope, 32
    graduating diaphragm, 32
Compressorium, 41
Condensers, 33
Condensing lens, 37
Connective tissues, 192
Conchifera, 170
Correlation of force, 117
Corti's organ, 223
Crystalline forms, 86, 114
Crystallization of salts, 100
Crystalloid, 66
Cryptogamia, 152
Crustacea, 172
Cyclosis, 129
Cystine, 241

Dammar mounting, 79
Darker's selenite stage, 44
Dark-ground illumination, 35
Decomposition of blood, 227
Definition, 54
Degeneration, fatty, 230
Dentine, 196
Desmidiaceæ, 140
Development of tissues, 201
    of fungi, 137
Diabetic sugar, 237
Diaphragm, 32
Diatom markings, 56
Diatoms, 94, 141
    classification of, 142
Diatomaceous earth, 94
Diffraction of light, 54
Discrimination of forms, 127
Dotted ducts, 131
Double refraction, 91
Dry mounting, 78
Dujardin, 118

Early microscopists, 20
Echinoderms, 167
Educational microscopes, 30
Epithelium, 190
Elementary unit in biology, 117
Epiphytes, 242
Epizoa, 243
Equisetaceæ, 155
Equisetum, 132

Embryo, sections of, 205
Embryonic development, 201
Entomostraca, 173
Entozoa, 171, 243
Enamel, 192
Eozoon, 84, 97
Errors of interpretation, 52
    from refraction, etc., 52
Examination, methods of, 58
    of minerals, 85
    of higher plants, 155
Eyes, care of, 49
    of insects, 176
Eye-pieces, 26

Fatty tissue, 195
    degeneration, 230
Feet of insects, 177
Ferns, 155
Fission of cells, 125
Fixed oil in plants, 133
Flatness of field, 55
Fluid media, 66
    mounting, 80
    cavities in minerals, 89
Flowers, 157
Focal distance, 22
Formed material, 118, 128
Forms of vegetable cells, 134
Formations, pathological, 232
Foraminifera, 95, 159
Fossil plants, 93
Fraunhofer's lines, 44, 101
Frog-plate, 41
Fungi, 134, 138

Gas chamber, 42
Gasteropoda, 170
Gemmation, 125
Generative organs, 213
Germ-cell, 125
Germinal matter, 118, 122
Geology, microscope in, 92
Glass-covers, 77
Glandular fibres, 131
    tissues, 200
Graduating diaphragm, 33
Grammatophora test, 57
Gum, 134

Hair, 191
    of insects, 175
Haller, 118
Hardening tissues, 61
Heart, 208
Hepaticæ, 154
Herapathite, 113
Herschel's doublet, 22
High powers, 55

INDEX. 257

Hipparchia Janira, 56
Histo-chemistry, 185
Histology, 182
Histologists, early, 20
Histological structures, 185
Holothuriæ, 168
Hunt, Dr., on staining, 153
Hydrozoa, 165

Imbedding tissues, 62
Immersion lenses, 25, 51
Indifferent fluids, 66
Infiltration, 228
Infusoria, 160
    families of, 163
Injecting, 64
    syringe, 64
    material, 65, 70
Insects, 174
    scales of, 175
    hairs of, 175
    eyes of, 176
    mouths of, 176
    feet of, 177
    anatomy of, 177
    changes of, 126
Intestinal canal, 206
Interpretation, errors of, 52
Inverted microscope, 106
Invertebrata, classes of, 180
Iod-serum, 66
Iris diaphragm, 33

Kellner's eye-piece, 26
Kidney, 211

Labyrinthodon, 97
Laticiferous vessels, 131
Lamps, 50
Leaves of plants, 157
Lasso-cells, 165
Lenses, 21
Lernœa, 173
Leptothrix, 136
Leucocytes, 189
Lieberkuhn, 37
Light for microscopes, 50
Life, theories of, 116
Litmus-paper, 107
Living bodies, element of, 117
Ligneous tissue, 130
Lichens, 154
Liver, 210
Locomotive organs, 215
Low powers, 55
Lymph, 189
Lymphatics, 207
Luteine spectra, 105

Magnifying power, 22
Max Schultze, 118
Measuring objects, 38
Medium powers, 55
Metric system, 39
Medicine, microscope in, 226
Metamorphosis, 126
Microscope, compound, 23
    mechanism of, 27
    in the arts, 17
    in commerce, 18
    in agriculture, 18
Micro-spectroscope, 44
Micro-mineralogy, 84
Micro-chemistry, 98
Micro-chemical analysis, 106
Microscopic slides, 77
Microzymes, 135
Milky juice in plants, 131
Mildew, 136
Minute lenses, 21
    dissection, 59
Micro-gonidia, 152
Micrometer, 38
Moist-chamber, 41
Moller's test-plate, 57
Mounting objects, 76
Molecular movements, 120
    coalescence, 138
Monera, 158
Mosses, 154
Morbid actions, 227
Morphological products, 183
Motile cells, 120
Mounting in balsam, 79
Mucus, 190
Mucoid softening, 231
Muller's eye-fluid, 68
Muscardine, 136
Muscle, 197
Mycelia of fungi, 137

Nachet's inverted microscope, 32
    prism, 35
Navicula rhomboides, 56
Nerve tissue, 198
    cells, 199
    preparations, 217
Nichol's prism, 43
Nobert's illuminator, 35
    test, 56
Nose-piece, 47
Nostochinæ, 151
Nucleus of cells, 119
Nutritive organs, 206

Oberhauser's drawing apparatus, 39
Objects of microscopy, 58
Object glasses, 25
    finders, 48

Oblique illumination, 34
Ocular micrometer, 38
Opaque injections, 65
    objects, 77
Opticians, list of, 27
Orbitolites, 159
Organs of touch, 218
    of taste, 218
    of smell, 218
    of sight, 219
    of hearing, 222
Origin of rocks, 92
Organic principles, 184
Oolites, 90
Otoliths, 222
Oscillatoriæ, 151
Oxalate of lime, 241

Pabulum, 118, 183
Paleontology, 97
Palmellaceæ, 139
Parasites, human, 242
Parthenogenesis, 126
Pathology, microscope in, 226
Pathological formations, 232
Parabolic illuminator, 36
    speculum, 37
Pettenkofer's test, 102
Penetration, 55
Periscopic eye-piece, 26
Preserving objects, 76
Pleurosigma test, 58
Pigott's, Dr. R., eye-piece, 26
Pigmentation, 229
Phosphates, 240
Physical movements, 53
Photography, microscopic, 48
Phenomena of cells, 120
Polariscope, 42, 99
    in mineralogy, 90
Polycystina, 95, 160
Polymorphism, 136
Polyps, 164
Polyzoa, 168
Popular microscopes, 30
Porifera, 160
Preparation, 58
    by teasing, 61
    by section, 61
    of minerals, 84
    of crystals, 85, 99
    in viscid media, 65
    of vegetables, 156
    embryonic, 204
Preservative fluids, 73
Preserving objects, 76
Primordial utricle, 129
Protoplasm, 118
Progressive force, 130
Prothallium of ferns, 126
Prism for oblique light, 35

Protophytes, 129
Protozoa, 158
Pus, 180

Raphides, 132
Reade's condenser, 34
Recklinghausen's moist-chamber, 41
Red blood-corpuscles, 186
Reproduction, 125
Resolution, 55
Resin, 133
Reticulated vessels, 131
Respiratory organs, 213
Retina, 221
Rhodospermeæ, 153
Rhizopods, 158
Rotatoria, 163

Sarcode, 118
Salivary glands, 209
    corpuscles, 189
Schleiden and Schwann, 118
Sclerogen, 130
Schultze's warm stage, 42
Scales of Lepidoptera, 175
Section-cutter, 63
Secretory organs, 208
Sections of hard tissues, 62
Sensory organs, 218
Selenite stage, 44
Seeds, 157
Shadbolt's turntable, 78
Shell structure, 170
Silica in plants, 131
Simple tissues, 186
Sœmmering's steel disk, 40
Spectra of substances, 102
Sperm-cell, 125
Spectrum analysis, 101
Spiral movements, 130
    vessels, 131
Spot lens, 36
Sphagnum, 155
Spherical aberration, 22
Sphæroplea, 152
Sponges, 160
Spontaneous generation, 125
Stage micrometer, 38
Staining cells, 122
    tissues, 63
    fluids, 69
Starch, 132
Stems of plants, 156
Still cells, 140
Students' microscopes, 28
Stricker's gas chamber, 41
Sublimation, 99
Surirella gemma, 56
Stomata, 157
Sputa, 243

INDEX.

Sweat glands, 209
Sympathetic nerve, 215

Table. 50
Tapeworm. 171
Tactile papillæ, 218
Tests, microscopic, 54
    micro-chemical, 106
    for alkalies, 108
    for acids, 109
    for oxides, 110
    for blood, 102
Theology and the microscope, 19
Thin cells, 77
Tissue elements, 185
Torula, 135
Tow net, 82
Transformations, 126
Transparent injections, 72
Triple phosphates, 240
Tumors, 232
Turntable, 78
Tunicata, 169

Uric acid, 240
Urinary deposits, 238
Ulvaceæ, 151

Varieties of bioplasm, 123
Vascular tissue, 201
Valentine's knife, 62
Vernier, 43
Vegetable cells, 128
Vibriones, 135
Vinegar eels, 171
Vorticella, 161
Volatile oil, 133
Volvox, 140

Warm stage, 42
Wandering cells, 121
Water bears, 164
    flens, 173
Wenham's prism, 30
Webster condenser, 34
White blood-cells, 188
Wollaston's doublet, 23

Xanthidia, 96

Zentmayer's microscope, 30
Zoology, microscope in, 158
Zoophytes, 164
Zygnema, 151

DIRECTIONS FOR THE PLATES.

Plate	1	to face page	56.
"	2	"	92.
"	3	"	112.
"	4	"	114.
"	5	"	120.
"	6	"	130.
"	7	"	132.
"	8	"	134.
"	9	"	136.
"	10	"	140.
"	11	"	154.
"	12	"	160.
"	13	"	167.
"	14	"	166.
Plate	15	to face page	168.
"	16	"	170.
"	17	"	176.
"	18	"	188.
"	19	"	192.
"	20	"	196.
"	21	"	200.
"	22	"	208.
"	23	"	210.
"	24	"	222.
"	25	"	234.
"	26	"	238.
"	27	"	240.

www.ingramcontent.com/pod-product-compliance
Lightning Source LLC
Chambersburg PA
CBHW021954220426
43663CB00007B/805